D0849061

The Field of Geography

General Editors: W. B. Morgan and J. C. Pugh

Plant Geography

Martin C. Kellman

Plant Geography

Methuen & Co Ltd

First published 1975 by Methuen & Co Ltd
11 New Fetter Lane, London EC4P 4EE
© *1975 Martin C. Kellman*
Typeset in Great Britain by
Preface Limited, Salisbury, Wilts
and printed in Great Britain
at the University Printing House, Cambridge

ISBN 0 416 81260 0 *hardback*
ISBN 0 416 81270 8 *paperback*

To Tila

Contents

The Field of Geography

Progress in modern geography has brought rapid changes in course work. At the same time the considerable increase in students at colleges and universities has brought a heavy and sometimes intolerable demand on library resources. The need for cheap textbooks introducing techniques, concepts and principles in the many divisions of the subject is growing and is likely to continue to do so. Much post-school teaching is hierarchical, treating the subject at progressively more specialized levels. This series provides textbooks to serve the hierarchy and to provide therefore for a variety of needs. In consequence some of the books may appear to overlap, treating in part of similar principles or problems, but at different levels of generalization. However, it is not our intention to produce a series of exclusive works, the collection of which will provide the reader with a 'complete geography', but rather to serve the needs of today's geography students who mostly require some common general basis together with a selection of specialized studies.

Between the 'old' and the 'new' geographies there is no clear division. There is instead a wide spectrum of ideas and opinions concerning the development of teaching in geography. We hope to show something of that spectrum in the series, but necessarily its existence must create differences of treatment as between authors. There is no general series view or theme. Each book is the product of its author's opinions and must stand on its own merits.

University of London,　　　　　　　　　　　　　　　　**W. B. Morgan**
King's College　　　　　　　　　　　　　　　　　　　　**J. C. Pugh**
August 1971

Figures

Tables

Plates

Preface

In writing this book on plant geography, I have had in mind those geography students whose interest in the landscape has extended to things biological. Biogeography, after a long absence from formal geography curricula, has undergone a recent revival. However, there is, as yet, no substantive body of literature in 'geographical' biogeography, and no clear conceptual framework within which the student may function. Under these conditions, the student is usually cast adrift in the literature of biological ecology and biogeography, with few guiding principles and little basic biological knowledge. In this book I have attempted to provide a conceptual framework with which such a student may approach plant geography, together with sufficient introductory plant biology to allow him to do so.

The result has been a generally two-tiered treatment of topics within the book: some basic factual information followed by a discussion of conceptual issues. The latter are often controversial in plant geography, and it is inevitable that my own opinions and predilections have emerged here. However, I hope that students will not treat these opinions as accepted dogma, but rather as indications that the field remains conceptually stimulating and open to continual reinterpretation. If students emerge from reading this book suspicious about most of what they read in the plant geographic literature, I will have achieved one of my main objectives. It has not been possible, within the scope of the book, to treat comprehensively all subject matter that may be placed beneath the rubric of plant geography. Consequently, I have concentrated upon those topics that I think are fundamental to an understanding of the field, while attempting to mention, wherever possible, other less central topics that nevertheless form research frontiers.

Some readers may be surprised that I have not chosen to open this book with a discussion of the ecosystem concept that is much in vogue today. This is based upon my belief, expanded upon in chapter 9, that the concept is a far more complex and intractable topic, and one that is far less useful, than is commonly realized. The processes of organic evolution have produced, in the biosphere, a phenomenon of great complexity, and one that does not readily conform only to the laws of thermodynamics. Thus, while I am sympathetic to those physical geographers who see in the concept a means of integrating animate and inanimate worlds, I believe that attempts to apply this, without first appreciating the complexity of its biological component, can have unfortunate results.

I acknowledge with thanks those who originally encouraged my interest in the plant landscape: Kenneth Hare, Paul Maycock, Jonathan Sauer and Donald Walker. I am also indebted to the many students who, willingly or otherwise, have been the subjects upon whom I have developed the ideas expressed here. Without the encouragement and assistance of my wife, Tila, this book could not have been written. She has read the manuscript many times and made invaluable improvements in style and content.

Belize, Central America
January 1974   M.C.K.

Acknowledgements

The author and publisher would like to thank the following for permission to reproduce copyright material:

P. G. Haddock, J. Walters and A. Kozak for fig. 2.1
Cambridge University Press for table 3.1
McGraw-Hill Book Company for fig. 3.2
British Columbia Forest Service for fig. 4.1
Duke University Press for figs. 5.1, 7.3 and 8.1
Australian National University Press for fig. 6.2
The Editor-in-Chief of the Commonwealth Scientific and Industrial Research Organization for fig. 7.1
C. D. Adams for fig 7.7 and table 10.1
Blackwell Scientific Publications Ltd for fig. 7.8
University of Chicago Press for fig. 8.8

1 Introduction

The scope of plant geography

Plant geography treats the plant cover of the earth's surface as an object of study. Within this broad field, many divergent approaches have been taken. In this book, the geographic theme of spatial relationships is adopted as the principal approach, although other themes are treated in the later chapters. The major portion of the book addresses itself to distributional problems of the earth's plant cover taken either as a whole (the geography of vegetation), or as separate components (the geography of species). The scales of distributional problems treated are without restraint, varying from global patterns at the smallest scale, to micro patterns of a scale governed by individual plant sizes. Although plant geography theoretically encompasses all classes of plant from the most primitive thallophyte to the most advanced angiosperm, it is seldom practicable, because of differing life cycles and physiologies, to treat adequately all such classes in a single work. Consequently, attention will be focused here on the higher (seed) plants which comprise the vast bulk of the earth's plant cover. However, many of the concepts enunciated are also applicable to lower plants.

The emergence of plant geography as a formal discipline lies in the works of Alexander von Humboldt, the great German geographer of the nineteenth century (Humboldt 1807). In his travels throughout Europe and Latin America, he made extensive collections of plant specimens and recognized the general correlative tendency between plant form and environment (mainly climate), both latitudinally and altitudinally. Thus he initiated the 'structural' approach to plant geography, a field which was developed by Warming (1909), Raunkiaer (1934) and others.

Twentieth century developments in plant geography have represented a progressive divergence in approach. On the one hand, some plant taxonomists have addressed themselves to distributional problems of plant

1

taxa (species, genera or families) or the floras of selected areas, leading to an approach usually referred to as floristic plant geography. The problems treated in this approach have usually been of a global or continental scale with explanations sought in historical factors such as rates of speciation, long range dispersal and continental drift. Studies emphasizing environmental explanations of plant distributions, which may be termed ecological plant geography, have developed under the aegis of plant ecology. The distributional problems treated have tended to be of a larger scale and historical factors have tended to be bypassed in favour of environmental explanations. Distributional problems of individual species have been treated as part of autecology (the environmental relations of the individual plant) while those involving vegetation have been treated as part of synecology (the study of plant communities).

Although this methodological distinction persists in much of the present literature on plant geography, it is in many ways unjustified. Most patterns in the earth's plant cover reflect both the effects of environment and a variety of historical events: rarely will one or the other be exclusively operative. Indeed, the assessment of the relative importance of these two groups of factors is often a fundamental task of the plant geographer, and one which cannot be circumvented by methodological arguments. In this work, such a synthetic approach is adopted in the hope that concepts can be enunciated which will allow plant geography, long fragmented among several subdisciplines, to re-emerge as a cohesive field of study.

The units of study in plant geography

As a phenomenon of great complexity, possessing many discrete attributes, the earth's plant cover is not amenable to any form of comprehensive geographical analysis. Instead, attributes which possess common properties must be selected as objects of study. In this book, two fundamental attributes of this plant cover are selected as the main units of study: the identifiable plant type, or species, and spatial assemblages of plant species. The species concept will be further treated in chapter 2, while the concept of plant species assemblages as objects of study is discussed in chapter 5.

In chapters 3 and 4, processes responsible for species' distributions are treated: chapter 3 discusses the environmental controls on plant species distributions while in chapter 4 the concept of site accessibility and processes of plant mobility are examined. In chapters 6 and 7 techniques of data gathering and data analysis in studies of vegetation are treated.

In chapters 8, 9 and 10, other geographical themes in so far as they apply to the plant landscape are discussed. The historical theme, strong in much of the geographical literature, is examined in chapter 9 in relation to terrestrial vegetation. The role of plants in matter and energy cycling through ecosystems is discussed in chapter 9. Chapter 10 focuses on the anthropic theme in geography, discussing man's role in transforming the pristine plant cover of the earth. In a postscript, some concluding thoughts on the future of plant geography are presented.

Distributional concepts in plant geography

Several general concepts in plant geography may be introduced at the outset. These apply primarily to distributions of species, but they can be extended, with slight modification, to vegetation distributions.

The first concept concerns the *range* of a plant species. The species is an abstraction synthesized from common attributes possessed by a number

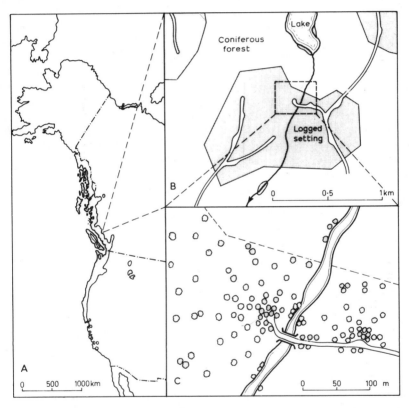

1.1 The range of red alder (*Alnus rubra*) at three scales. Maps B and C diagrammatic. At a continental scale (A), the species range appears to be correlated with a humid oceanic climate. (However, note disjunct interior and southern populations.) At a more local scale (B), the species range appears to correspond closely with logged areas. At a very detailed scale (C), the locations of individual trees are found to be confined to sites of severe disturbance within logged setting, especially logging roads, loading sites and drag-trails. Within undisturbed forest, the species occurs as scattered individual trees along stream banks and lake shores.

of discrete plants (chapter 2). Consequently, the range of a species cannot be envisaged as a continuous phenomenon occupying part of the earth's surface, but it is rather a collection of points, each representing an individual plant, spread over some area. The proximity of the points, and the scale at which the distribution is being considered, will determine the form of an envelope curve drawn about these points to specify the species range. Consequently, all species ranges represent spatial generalizations and possess internal discontinuities (fig. 1.1).

A second concept concerns the *potential range* of a species as set by physical and biotic factors and the internal physiological properties of the constituent plants. This may be conceived as comprising all areas habitable to members of the species should they be able to get there, and is normally far in excess of the actual area occupied by the species.

The *actual range* within this potential is determined by previous evolutionary events, environmental changes and the ability of the species to migrate. Consequently, each portion of the earth's surface may be envisaged as being able or unable to receive diaspores of certain species: the concept of *site accessibility* (Kellman 1970c). However, because innumerable factors and processes combine to determine whether or not a site is able to receive propagules of some species at any specified time, a fourth concept, that of *probability*, becomes fundamental to understanding the earth's plant cover. Thus each portion of the earth's surface, in addition to possessing a set of environmental conditions which determine whether a plant species or assemblage can survive there, also possesses at any instant in time, a certain probability that the species or assemblage can arrive there.

Part 1/The geography of plant species

2 The plant species and its ecological properties

The species concept

Although each plant is in some way morphologically unique, careful examination of a large number of plants reveals that these tend to fall into a number of fairly distinct categories on the basis of external appearance. For example, each sugar maple tree differs from all other sugar maples in details such as branching pattern, canopy shape and overall size. However, a close examination of large numbers of sugar maples reveals that there are far more similarities between them than there are differences: each has a simple leaf with three to five lobes arranged oppositely on the stem, fissured bark when mature and almost identical reproductive structures. Together they form a morphologically homogeneous group which, although possessing some internal differences, is fairly distinct from other comparable groups.

Such morphologically homogenous groups have been termed 'species' and are conventionally named with a Latin bi-nomial. Sugar maple is designated *Acer saccharum*, indicating both its distinctness as a species and its membership in the maple genus, *Acer*. Careful field observations and experimentation will reveal that the morphological homogeneity characterizing the species is paralleled by similarities in physiological properties, notably environmental requirements. Thus the species is an abstraction comprised of plants possessing certain common attributes of form and function.

The basis of homogeneity within a plant species is interbreeding through cross-pollination. Indeed, an ability to interbreed is a frequently used functional definition of the species (Löve 1962). This ensures continual interchange of genes, which control the form and physiology of the plant, and ensures that new genetic material, unless grossly incompatible with that existing, is incorporated into the genetic pool of the species.

7

A number of mechanisms are involved in the interchange of genes in an interbreeding population. Chief among these are the random sorting and recombination of parental chromosome pairs in the reproductive process, and changes within chromosomes due to crossing over of genetic material between pairs. These processes are treated fully in introductory texts on genetics and cannot be discussed in detail here. However, of importance in plant geography is the consequence that genetic recombination allows not only relative homogenization within an interbreeding population, but also some random genetic variability within it. This ensures a breadth of response to environmental conditions which may be of survival value to the species in the face of environmental fluctuations. A species population faced with some new environmental stress, such as a prolonged dry spell, may survive this period if sufficient genetic variability exists within it to ensure that some gene combinations impart drought resistance. In contrast, a genetically identical population (which may be produced by prolonged vegetative propagation from an isolated plant) would be faced with extinction if the single gene combination present did not impart drought resistance.

Origin of new species

New plant species seldom appear rapidly, save in some cases of hybridization and polyploidy. Rather, the origin of new species lies in the gradual transformation of existing species populations into new forms or the development of offshoots of these due to isolation. Isolation within the species may be the product of several factors prohibiting cross-pollination: physical isolation in a widespread species or biological isolation due to differing flowering times or the absence of a suitable cross-pollinating agent.

The appearance of a new species also depends upon the development of new genetic material within the population, whether this be an existing species or isolated portion of one. Without this, existing genetic material can only continue to be recombined in a limited number of ways. New genetic material is produced by a mutation, which may be defined as a spontaneous heritable change in genetic material. Mutations are random variations that can be produced artificially by ionizing radiation and certain chemicals, and are undirected by environmental conditions. The result of the mutation is for the plant affected to have some actual or potential new property of form or physiology. If the new property is lethal to the plant or its progeny it will ultimately be eliminated from the gene pool. If advantageous, the plant will survive and may prosper relative to others in the population, ensuring efficient reproduction and transmission of the trait to the gene pool. Thus, environment may be conceived of as a 'sifter', selecting from random mutations presented to it those that are advantageous, and eliminating those that are not. However, mutations which produce non-adaptive traits that are linked genetically or developmentally to some adaptive trait, may also survive the environmental sifting process and be incorporated (at least temporarily) into the gene pool. Consequently, it is unwise to suppose that all properties of a plant are necessarily adaptive: many may represent such artifacts. The implications of this phenomenon to the problem of environment and life form are discussed below.

The result of continual incorporation of acceptable genetic changes in an interbreeding species population is its slow transformation toward new forms and physiologies. Under stable environmental conditions, such evolution may be expected to lead towards increased competitive efficiency of the genotype until the available gene complex is stabilized at its optimal level for that environment. In contrast, where some progressive environmental alteration is in progress, such change may be expected to be channelled toward producing genotypes better adapted to the new conditions. In a species possessing no isolated subpopulations, such changes will result in its gradual transformation into new forms by 'linear' evolution. In contrast, where isolation develops within a species population, such changes will lead to slow divergence of the isolated populations and the eventual emergence of two or more distinct species.

Consequently, the species should not be regarded as a static unit but one which through time is modified into new forms, and which may be fragmented due to the vagaries of environmental change and plant migration. Rates of evolution appear to have been highly variable between different plant groups, as witnessed by the survival of ancient types such as the genus *Gingko* to the present, contrasted with the explosive evolution of such modern genera as *Eucalyptus* in Australia during the Tertiary.

Spatial variability within species

As a unit whose integrity depends upon contiguity of member plants and interbreeding between these, it is not surprising that one finds, in many widespread species, appreciable variability. Because the sifting effect of the environment impinges most directly upon a plant's physiological properties, determining survival, evolutionary changes in these properties often precede morphological changes. Consequently, in a widespread species, exposed to a variety of differing environmental stresses in different parts of its range, one often finds consistent physiological differences representing divergent evolutionary adaptation to these differing environments. The same phenomenon may also exist in less widespread species which are effectively separated into different gene pools due to internal or external isolating mechanisms. This phenomenon was first noted by Turesson (1922) who termed the different physiological races *ecotypes*. The existence of ecotypes is now widely documented in many species (fig. 2.1). In widespread species, whose ranges are relatively continuous, clinal variations in physiological properties, rather than

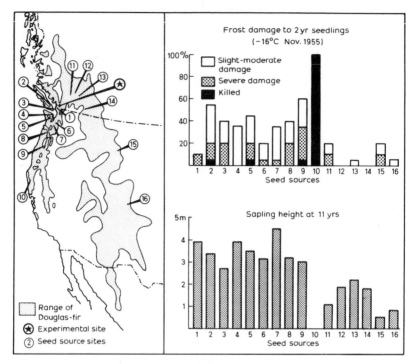

2.1 Ecotypic differences in Douglas-fir (*Pseudotsuga menziesii*) of different provenance. Note contrasts between the performance of interior and coastal provenances and the frost sensitivity of the California provenance. Source: Haddock *et al.* (1967).

distinct disjunctions, have been demonstrated (e.g. Olmsted 1944). This phenomenon, termed ecoclinal variation, is produced by varying degrees of isolation, and hence varying frequencies of gene interchange, throughout the species range.

Ecotypic differences within a species may be accompanied by slight morphological differences and hence be visually detectable. However, frequently a uniform morphology prevails throughout the species, requiring experimental demonstration of ecotypes. Such ecotypic variability imposes strictures on the concept of the species as a homogenous unit of study in plant geography, for it is seldom practicable to conduct the lengthy experimental work necessary to establish its existence. It requires that caution be exercised when generalizing about the environmental controls on some species, for these may differ in different parts of its range. However, despite these constraints, the species remains a useful unit of study in plant geography, because internal differences seldom exceed interspecific differences. Consequently, it can be regarded as a unit of *relative* homogeneity.

The phylogenetic system

The diversity of plant life present on the earth's surface represents the temporary endpoints of an ongoing evolutionary process, and considerable attention has been paid to tracing the evolutionary lines of development that have led to this. This has normally involved the development of a classificatory hierarchy thought to reflect the anastomozing evolutionary 'tree'. This has been termed a phylogenetic or 'natural' classification.

At the lowest level in such classifications, species showing internal morphological differences, but still thought to be potentially interfertile have been separated into subspecies or varieties. Thus, Douglas-fir (*Pseudotsuga menziesii*) in the Pacific Northwest has, in recent years, been separated into interior and coastal varieties (var. *glauca* and var. *menziesii*) on the basis of slight morphological differences. At higher levels in such classifications, species thought to be closely allied are grouped into genera. These in turn are grouped into families which themselves are grouped into a limited number of orders. At the highest level in the hierarchy, similar orders are grouped into a few large phyla, ranging from the primitive Thallophyta to the advanced seed-bearing Spermatophyta.

Because such hierarchies represent hypothetical summaries rather than proven facts, there is considerable variation between systems. The lower levels of taxonomic systems are continually being re-ordered as further taxonomic work is completed on specific groups. The considerable modification which species groups may undergo at the hands of taxonomists over the years illustrates the variability which exists in these groups and the need for caution when generalizing about them.

Environment and life form

As early as 1807, Humboldt had recognized the tendency for certain plant physiognomies (and hence vegetation structures) to be concentrated in specific environments. However, the explanation of such correlations remained obscure until the exposition of natural selection and evolution by Darwin in 1859 (Darwin 1859). Darwin's work set in motion a period of widespread attempts to find adaptive value in all features of organisms. In botany, this led to 'explanations' of the adaptive value of various plant life forms (Schimper 1903; Warming 1909).

While the observations of these early workers on environment/life form correlations were undoubtedly accurate, the exclusively adaptive explanations offered must be modified in light of more widespread field data and developments in the fields of evolutionary theory and genetics. Observations from widely separated parts of the globe, possessing floras of very different evolutionary history and gene pools, indicate the very different life forms that can evolve in response to similar environmental opportunities. Nowhere is this better illustrated than in Australia, where near-endemic genera such as *Eucalyptus* have evolved life forms in arid environments very different from those found in comparable environments in the northern hemisphere. This has resulted in vegetation types which bear scant resemblance to their northern temperate counterparts (Beadle 1951). Analagous conditions exist locally also, where the wide variety of life forms existing in the same environment bespeaks the importance of genetic ingredients in shaping this. Few life forms could differ more than the succulent cactus, the ephemeral herb or the seasonally deciduous shrub, yet all survive successfully in the arid regions of North America.

Furthermore, adaptations allowing a plant to survive in some environments are not necessarily morphological. Physiological adaptations are the prime requisite for survival, and morphological manifestations of these need not necessarily exist. In addition, the possibility that non-adaptive morphological features may exist in a plant as the result of developmental artifacts or gene linkage, also requires caution to be exercised when explaining plant life forms environmentally.

Despite the divergences introduced by the processes mentioned above, there are, to be sure, convergent tendencies in the evolution of plant forms among various taxa, especially in areas such as the tropics that appear to have been relatively stable for long periods of time. Nevertheless, the complicating effects of evolutionary history and ecotypic variability within plant species, and the burial of explanations in evolutionary history, suggest that further attempts to 'explain' existing life forms will be an unproductive exercise. Indeed, there are theoretical reasons to suppose that inter-plant competition may be the main shaper of life form evolution. Competition, as will be discussed below, is an all-pervasive phenomenon because few plants grow in isolation. Consequently, the impact of environmental conditions on a plant are usually qualified by the competitive effects of adjacent plants. Therefore, one may expect selective pressures to favour forms permitting effective competition with other plants to which a species has been exposed for prolonged periods, and the evolution of similar or equivalent forms in different taxa. This process may be expected to lead to a competitively balanced assemblage of genotypes (and possibly life forms) occupying a given environment rather than a population which has evolved perfect efficiency there.

Ecological properties of a species

The response of plant species to environmental conditions can be resolved into a number of general concepts. These include tolerance, ecological optima and ecological amplitudes.

If the performance of a plant species is examined along some environmental gradient to which it is sensitive, such as moisture or light, certain phenomena can be observed. Most significantly, beyond certain upper and lower threshold values of the gradient, the species is unable to survive. These values have been termed the tolerance limits of the species, and the intervening range its *range of tolerance*. Survival of a plant may include its ability to live for any period of time, but may be more narrowly defined as the ability of a plant to successfully complete all phases of its life cycle. Under natural conditions a plant species must be capable of completing all phases of its life cycle in a given area if it is to persist there for a prolonged period (Pelton 1953). However, the survival of artificially introduced species in marginal environments for short periods and the non-reproduction of many species populations under natural conditions, indicate that not all phases of a species' life cycle have similar environmental requirements. Individual phases, such as vegetative growth, may have a far wider tolerance range than all phases in combination.

A second response that will be observed is that plants of the species being treated will not grow with equal vigour in all parts of their tolerance range. Within the central part of their range the plants will normally grow most vigorously, this decreasing progressively towards the tolerance limits where it becomes zero. The portion of the range where vigour is greatest has been termed the *ecological optimum* of that species (fig. 2.2).

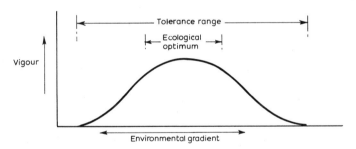

2.2   Tolerance range and ecological optimum of a plant species.

If the performance of the same species is examined along other environmental gradients, comparable responses will be observed to all of those variables to which it is sensitive. These separate responses would include a great variety of ranges, ecological optima and vigour within these. The sum of tolerance ranges along all gradients to which the species is sensitive has been termed the *ecological amplitude* of the species. Within this summed hyperspace will exist a more restricted zone (or zones) of greatest vigour; this may be termed the optimum ecological amplitude of the species. Although the great number of environmental variables to which a plant species may be sensitive makes it impossible in practice fully to define the ecological amplitude of a plant species, the concept provides a useful working approximation within any set of environmental variables being treated.

If comparable observations were to be extended to other species, it would be found that each would respond in a way which was comparable, although differing in detail. Species would differ in their sensitivity to certain gradients: different species would possess differing tolerance ranges and ecological optima. In sum, each species would have a differing ecological amplitude. However, this would not preclude an overlap of ecological amplitudes; if this were so, no two species would ever grow together! Although the existence of similar ecological amplitudes in areas of great floristic richness, such as the tropical rainforest, is a question of considerable debate, the initial assumption of uniqueness in species' ecological amplitudes provides a useful preliminary hypothesis in plant geographic research. When combined with the concept of the plant species as a unit of *relative* homogeneity, species ecological amplitudes may be conceived of as comprising a family of somewhat differing ecotypic amplitudes, but with internal differences within these less pronounced than the interspecific differences between them (fig. 2.3).

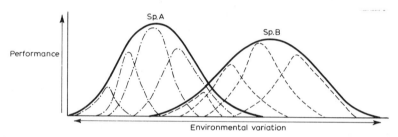

2.3   Inter- and intraspecific differences in performance of plant populations along an environmental gradient.

Competition

Competition occurs between two plants when both draw upon the same resource of which there is limited supply. The effect of such competition is the suppression or elimination of the less efficient utilizer of the resource. The resources may be any of those utilized in plant growth, such as light, soil moisture and soil nutrients. In arid environments, soil moisture usually becomes the most contested resource: only a low density of plant material can be supported in such environments and the widely spaced plants may compete vigorously with extensive root systems. In contrast, in moist habitats capable of supporting high plant densities, competition for the light available to the tighly packed above-ground plant tissue is normally paramount.

Competition may take place between individual plants of the same species if these grow in close proximity to each other, or between individuals of different species. Intraspecific competition is seen most clearly in the colonization of bare areas, where a single species may form pure stands whose density is dramatically reduced during maturation. For example, bare mineral soil in coastal British Columbia is often colonized by a pure stand of red alder (*Alnus rubra*). During the first year of colonization, densities of 75 seedlings per square metre may occur. These densities are halved within 3 years and further reduced to 1.89 saplings per square metre at 8 years and 0.03 trees per square metre at 26 years (pl. 1).

Because few plant species grow isolated from other species and because few environments are sufficiently bountiful to preclude competition, interspecific competition is an everpresent phenomenon in the earth's plant cover. Consequently, most plant species have available to them environments whose resources are far more circumscribed than were they to be grown in isolation. The result of this phenomenon is actual plant species' ranges that are far narrower than their full tolerance ranges. Many eastern North American boreal forest species can, if protected from competition, be grown successfully in the deciduous forest zone to the south, yet under natural conditions are unable to compete there successfully with the native deciduous forest tree species. Similarly,

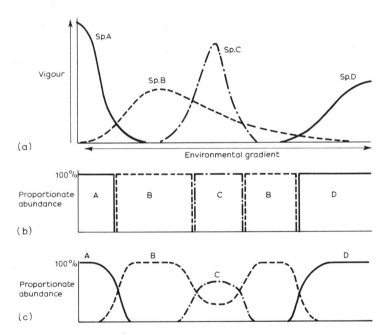

2.4   Performance of competing species. (a) Responses of species grown in isolation. (b) Theoretical field response when grown in competition. (c) Commonly observed field response (simplified). The theoretical expectation of competitive exclusion is usually modified by extraneous local environmental variation and genetic variability within the species populations, resulting in some species coexistence (cf. fig. 5.1).

botanical gardens containing large numbers of exotic species grown free of competition are rapidly impoverished by competition from aggressive local species if left untended. In this way, apparent environmental responses observed in the field may be appreciably modified versions of true responses. Species may be reduced in, or excluded from, parts (even optima) of their range by more efficient competitors, producing what appear to be narrowed, skewed, or polymodal response curves (fig. 2.4). In chapter 3 it will be shown that other forms of plant interaction also exist, whose manifestations are often very similar to those of competition.

3 Environmental control of plant species' distributions

The distinction between actual and potential ranges of plant species has been made earlier. This chapter treats those factors and processes responsible for setting potential ranges of species, while the role of plant mobility in determining the actual ranges within these is treated in the following chapter. Elements of both the 'physical' (or inanimate) environment and the 'biotic' environment are involved in setting potential ranges although, in detail, the effects of biotic interactions are most frequently exercised through alteration of the physical environment surrounding the affected plant. The role of biotic interactions in determining species ranges is also highly complex and variable in space and time because of the large number of different interacting organism combinations that may confront any individual of a plant species.

The concept of environment

The concept of environment has been extensively discussed by Mason and Langenheim (1957) who propose the following definition of the term:

> The environment of any organism is the class composed of the sum of those phenomena that enter a reaction system of the organism or otherwise directly impinge upon it to affect its mode of life at any time throughout its life cycle as ordered by the demands of the ontogeny of the organism or as ordered by any other condition of the organism that alters its environmental demands.

This provides a usefully organism-centred definition of the term 'environment', a term which has far too frequently been given a definition so wide as to be meaningless. It is common in the social sciences to use 'environment' to refer to any condition in which, or with which, the responding phenomenon exists, while a recent awakening of public interest in 'the environment' has witnessed the further extension of the

17

term's usage to cover virtually all phenomena of the real world. In contrast, Mason and Langenheim's definition provides a suitably clear and logical definition: phenomena which do not impinge upon the organism in any way are considered 'non-environment', while those which may at some future time fulfil this role are designated 'potential environment'.

Despite the value of this definition, its practical application faces several obstacles. It is seldom possible to identify precisely those factors impinging directly upon an organism due to collinearity between variables, difficulties of measurement and insufficient knowledge about the physiology of the organism. This has led some workers to propose abandonment of the search for cause and effect environmental relationships proposed by this definition in favour of a search for merely functional relationships (e.g. Major 1961). However, to abandon the search for causality implies abandoning the physiological basis of a plant's response to its environment, a procedure that could ultimately lead to a 'shotgun' approach to the search for environmental relationships. A suitable compromise would seem to be to invoke causality in formulating research designs and generating hypotheses, but to interpret correlations observed in the field cautiously, as reflecting relationships that are potentially only functional and not necessarily causal. This problem of interpreting field observations will be further treated when the field environment is discussed.

Dunbar (1968) has made a useful distinction between proximate and ultimate environmental effects. Proximate effects are those affecting the organism as it is presently evolved, while the ultimate effects comprise those operating in the past which have led to the evolution of present proximate responses. Although attention here will be focused upon proximate effects, it is important that the distinction between the two be borne in mind: there is a tendency in the plant geographic literature to interpret observed proximate responses ultimately, ignoring the fact that these are the syntheses of unknown past evolutionary events and may not represent optimal adaptation.

It is not possible to treat here comprehensively all environmental stimuli that may affect higher plants; several excellent treatments already exist to which the reader is referred (e.g. Daubenmire 1959; Russell 1961). Rather, the basic life cycle and environmental requirements of the autotrophic higher plant will be briefly sketched after which the conceptual and methodological problems of examining environmental control over species' distributions will be treated.

Ontogeny of the higher plant

Ontogeny refers to the entire developmental life cycle of an organism, here restricted to the seed-bearing autotrophic plant. If the seed stage is temporarily ignored, this comprises the following generalized sequence: germination, growth, reproduction, senescence and death. Germination comprises those processes by which a quiescent seed is transformed into an independent photosynthesizing juvenile plant. It normally involves stimulation of metabolic activity within the seed, emergence of the proto-root (radicle), proto-stem (hypocotyl) and proto-leaves (cotyledons) from the seed coat and eventual development of true organs. During this period the seedling is sustained by reserve materials contained within the seed endosperm or cotyledons. The environmental stimuli to seed germination are many and varied including moisture, temperature, light and atmospheric gases (Mayer and Poljakoff-Mayber 1963). Where the appropriate stimulus, or combination of stimuli, for a species' seed germination does not occur, the species will be effectively excluded from colonizing. Similarly, the delicacy of the emergent seedling makes it particularly susceptible to environmental stresses such as dessication and fungal infection. For this reason, the seeds of many species can be found germinating in habitats where ultimate establishment is never achieved.

Establishment of the juvenile plant with exhaustion of seed reserves brings with it the requirement for an adequate photosynthetic rate. Drawing upon environmental resources of soil moisture, atmospheric carbon dioxide and incident short-wave radiation, the plant must synthesize carbohydrates at a rate in excess of the rate of carbohydrate breakdown for respiration, if growth and development is to be ensured. This is often a critical period for the plant which, being small, is frequently growing in the dense shade of others. While the high carbon dioxide levels produced by plant litter decomposition in such habitats may partially compensate for this deficiency, there is normally a high mortality rate during this phase. Those juveniles which do survive this rigorously competitive stage grow eventually to enter a mature phase in which reproduction is achieved by flowering and, in some instances, vegetatively. Both of these processes in any species may have their own specialized environmental requirements which, if unfulfilled, will preclude continuance of the species beyond one generation. The final senescent stage of the life cycle is an ill-defined and poorly understood phase during which overall efficiency of the plant declines and the probability of infection by pathogenic organisms increases.

Photosynthesis in higher plants is parallelled by a second fundamental process to all terrestrial plant life: transpiration. The soil moisture required by the plant for photosynthesis represents only a minute proportion of the total needed, almost all of which is utilized in transpiration. In this process, moisture removed from the soil by the plants' roots is translocated up the stem via the xylem conducting tissue to leaves where it is evaporated into the atmosphere through pores called stomata, opening diurnally under a light stimulus. The ultimate adaptive value of this mechanism is unclear, although it effectively reduces the temperature of leaves exposed to full insolation, and may be important to mineral absorption and translocation within the plant. Whatever its ultimate adaptive value, transpiration imposes a set of environmental requirements and susceptibilities on the plant which is often critical to its survival. An adequate supply of utilizable moisture must be available to the plant's roots, while an increase in any atmospheric condition favouring increased evaporation (higher incident radiation, higher temperature, lower humidity, higher windspeed) can increase transpiration losses and promote dessication of the plant.

The ontogeny of the higher plant thus poses a series of environmental requirements which include both environmental resources utilized in growth, and other non-resource environmental conditions that may affect the plant's physiology at some stage of its life cycle. While these requirements are met directly by the abiotic environment, the process is frequently complicated by the interaction of other organisms, operating either directly (symbiosis, parasitism, predation), or indirectly (competition, allelopathy).

Abiotic effects

While it is theoretically possible to separate environmental effects exercised through varying environmental resources, from those involving only impingement on the plant's physiology, in practice, many environmental resources also exercise non-resource effects upon plants. Consequently, although environmental resources are first discussed below, mention is made, where appropriate, of other effects that these may have upon species' distributions.

Moisture

At global and continental scales, the availability of utilizable moisture is one of the most potent determiners of species' distributions. For most plants, only moisture in the soil is available for transpiration and growth. In soil that is not water saturated, moisture exists as films around soil particles, held there by electrostatic forces of attraction between the water molecules and the particle surface which exceed the gravitational forces promoting downward movement (fig. 3.1). Soil filled to saturation and then allowed to drain will rapidly lose all 'gravitational' water removed by gravitational force. In this free-drained state the soil is said to be at *field capacity*. Plant roots can draw upon this water provided that they can overcome the forces binding water molecules to the soil particles. As water is removed, moisture films become progressively thinner and water molecules more tightly bound to the soil particles (fig. 3.1). To remove more moisture, the plant must exercise ever more suction. At some point along this energy gradient, the plant is unable to do so any further, becomes dehydrated, and wilts.

Because of these properties of soil moisture, the critical fact of moisture supply to the plant is not merely the total amount present in the soil, but rather the free energy condition, or *tension,* of this. Soil moisture tensions can be measured directly at high moisture contents by using the

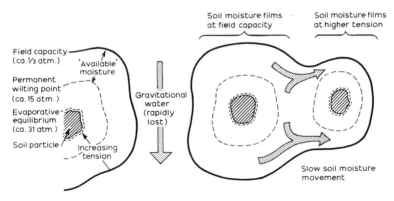

3.1   Conditions of soil moisture (diagrammatic).

direct sucking action of a mercury manometer connected to a porous cup sensor. At higher tensions, determination involves the differential expelling of moisture from a soil sample through a semipermeable membrane or ceramic plate under increased atmospheric pressure. Soil moisture tension measurements can be expressed as the length of a water column providing the appropriate suction, or the atmospheric pressure, measured in atmospheres or bars, necessary to expel the moisture through a semi-permeable medium. It can be shown experimentally that soil moisture at field capacity has a tension of approximately 1/3 atmosphere. The wilting point of many herbaceous plants approximates 15 atmospheres, although many desert perennials can withdraw moisture at far higher tensions, even down to the 31 atmospheres corresponding to the evaporative equilibrium with moisture in the soil atmosphere. While gravitational water moves rapidly downward in the soil, moisture at field capacity and higher tensions moves far more slowly from areas of low tension to areas of high tension until equilibrium is achieved. However, this rate of movement is insufficient to keep pace with the plant's utilization of moisture, requiring plants in areas of water deficit continually to expand their root network to new sources. Those species unable to do this cannot survive. In subhumid temperature grasslands, many grass species appear to have effectively compensated for moisture deficiencies by evolving the poten-tial for extensive root system ramification to new sources of soil moisture (fig. 3.2).

Although the effects of soil moisture on species' distributions are normally exercised through potential deficits, excessive soil moisture may have an indirect effect through reduction of root aeration. Different species have evolved differing tolerances to this phenomenon represented in extreme instances by such morphological features as root 'knees' in bald cypress (*Taxodium distichum*) and pneumatophores in black man-groves (*Avicennia* spp.) (pl. 2).

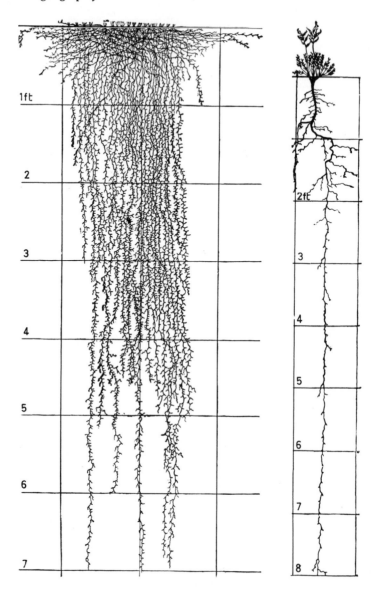

3.2 Root networks of Buffalo grass (*Buchloe dactyloides*) (left) and loco weed (*Oxytropis lamberti*) (right) in mixed prairie. Note the extensive root network of the grass in the upper soil horizon, contrasting with the single forb taproot that extends to ground water. Source: Weaver and Clements (1938). Copyright © 1938 J. E. Weaver and F. E. Clements. Used with permission of McGraw-Hill Book Company.

Soil nutrients

Higher plants require the following seventeen elements for growth and other metabolic processes: carbon, hydrogen, oxygen, nitrogen, phosphorous, potassium, calcium, magnesium, sulphur, iron, manganese, copper, boron, zinc, molybdenum, chlorine and cobalt. Carbon, hydrogen and oxygen are obtained from air and water, while all remaining elements must normally be derived from soil sources. Of these, manganese, copper, zinc, boron, molybdenum, chlorine and cobalt are required in such small quantities that they are not normally deficient save in soils of great age or derived from parent materials of aberrant chemical composition. These elements have been termed *trace elements*. Nutrient elements are absorbed in ionic form from the soil solution or, especially in the case of cations, by exchange reactions from their sites of adsorption on the surfaces of organic and mineral soil colloids. While the ions present in solution are readily available to plants, those adsorbed onto colloids are available only through expenditure of chemical energy for the exchange reaction. Species differ, not only in their total nutrient requirements, but also in their ability to extract adsorbed ions: many native species in the tropics flourish on soils where other species, notably cultigens, are unable to survive.

Soil nutrient measurements are an enigmatic subject due to the absence of any technique which measures the 'exchangeability' of adsorbed ions in a way analogous to soil moisture tension measurements. Thus, while the readily available ions in soil solution can be extracted by expelling the solution, those adsorbed onto colliods can only be partially removed by leaching with some arbitrarily selected chemical reagent. These determinations have, for obvious reasons, been keyed primarily to cultivated species and the most widely utilized reagent for extracting cations has been a normal solution of ammonium acetate. However, the results achieved with this and other extractants have not been notably successful in explaining the distribution of most wild species, save in instances where dramatic discontinuities in soil nutrient availabilities exist. Rather, meaningful tests of most wild species' reactions to soil fertility can only be achieved by experimental plantings of these species on different soils and trials using selective application of individual nutrient elements.

As in the case of soil moisture, excesses in the soil of nutrients and other non-required elements may become inhibiting to growth. This inhibition may be exercised through an excess of an absorbed element becoming toxic, such as aluminum under very acid conditions, or through a rise in the ionic concentration of the soil solution, thereby increasing the osmotic gradient that a plant must overcome if soil moisture is to be extracted by its roots. The latter condition is particularly common in the saline environment of sea coasts and a major hazard of irrigated farming in arid environments. Many salt tolerant species, such as mangroves, appear to overcome these high osmotic differentials by possessing cell sap of comparably high concentration.

Soil pH

The concentration of hydrogen ions in the soil solution, or its pH[1] is of special importance to plant nutrition. Although many plant species appear sensitive to soil pH, it has been shown that hydrogen ion concentration *per se,* at the levels normally found in soils, do not have any direct effect upon plants (Russell 1961). Rather, the effect is an indirect one, operating through the differential solubility of various required or toxic ions at various soil pH levels. An alternate use made of soil pH measurements has been to treat them as a crude indicator of soil fertility. The concentration of hydrogen ions in the soil solution is in a dynamic equilibrium with those in the exchange complex so that variations in the latter are reflected in the former. Parts of the cation exchange complex that are not filled by the main nutrient cations; Ca^{++}, Mg^{++}, K^+, Na^+, are filled with H^+ ions. Consequently, where the exchange complex remains constant, a change in hydrogen ion concentration gives some indication of changes in the total amount of other cations. However, the considerable variability in cation exchange capacity between soils and the instability of the organic colloid fraction of a soil's cation exchange capacity through time, makes this use of pH measurements very tenuous.

Insolation

Although variations in the income of short wave solar radiation used in photosynthesis are appreciable at a global scale, its most profound effect upon species' distributions is at a very localized scale where competition for light and shading by larger plants' foliage is a powerful determiner of species' distributions. Solar radiation is received at the outer surface of the earth's atmosphere at an intensity of approximately 2 cal./cm^2/min. and with a spectral composition as shown in fig. 3.3. Passage through the atmosphere depletes and distorts this spectrum by reflection, scattering and absorption, resulting in inputs at the earth's surface that vary continuously in both quantity and composition. The photosynthetic process does not utilize all wavelengths equally, being especially active in the 0.6—0.7 micron (red) and 0.4—0.5 micron (blue) wavebands. The intervening green waveband is little utilized and most is reflected, resulting in the characteristic colour of photosynthetic tissue. Other physiological responses are elicited by specific wavebands. These include chlorophyll formation, photoperiodism and seedling responses in the red wavelengths and phototropism in the blue wavelengths (Wassink 1953). Furthermore,

[1] The pH of a solution is defined as the logarithm of the reciprocal of the hydrogen ion concentration, expressed in grams per litre.

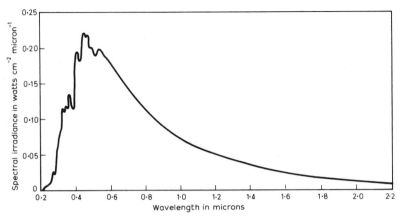

3.3   Spectral composition of extraterrestrial insolation. Source: Data from Johnson (1954).

radiation of all wavelengths, including longwave re-radiated radiation, affects plants through promoting transpiration.

Although global variations in solar radiation inputs occur, the very small percentage of this that is fixed in the photosynthetic process in all areas (approximating 1%), suggests that insolation *per se* is not a prime determiner of distributions at this scale. However, within stands of vegetation, great variations occur both in total input and in spectral quality. Total inputs of less than 1% of external values can be measured at the soil surface within forest stands, while the spectral quality of this may be distorted appreciably by the overhead canopy (Coombe 1957; Robertson 1966; Vézina and Boulter 1966). Under these conditions, variations in 'shade' insolation can have a profound effect upon local species distributions. A shade adapted flora, unable to withstand high insolation levels and associated environmental changes, frequently occupies the floor of many forests, while other species, unable to germinate or grow at such insolation levels, are excluded.

Atmospheric gases

Atmospheric carbon dioxide, as a freely diffusing gas, seldom shows marked changes in concentration over large areas. However, within stands of vegetation, the release of carbon dioxide from decomposing plant residues on stand floors, and the low rates of air movement at such sites, may lead to higher concentrations there. Because most plants are capable of high photosynthetic rates when exposed to levels greater than 0.03% occurring in the free atmosphere, this may partially compensate for the low insolation levels in such habitats. Conversely, carbon dioxide concentrations may rise to toxic levels (over approximately 10%) in poorly aerated soil.

Until recently the influence on plant distributions of atmospheric gases, other than carbon dioxide, has been slight. Oxygen, required for respiration, is normally in sufficient supply in the free atmosphere of which it comprises 21%. Only under conditions of soil waterlogging does an oxygen deficiency in the soil atmosphere have some influence on plant roots. However, the expansion of industrial activity in the twentieth century has resulted in a wide range of new gases being released into the atmosphere in high concentrations. Many of these are toxic to plants, and around urban and industrial centres, effects upon the vegetation have been profound (Sinclair 1969). One of the most toxic of these gases has been sulphur dioxide, released in large quantities during industrial smelting processes. While at low concentrations some species are capable of surviving, in areas that have been exposed to excessive fumigations, such as those surrounding Trail, British Columbia and Sudbury, Ontario, wholesale destruction of the entire vegetation cover has ensued (pl. 3).

Temperature

Of all non-resource environmental phenomena, temperature has the most profound effect upon plant distributions, especially at a global scale. Plants possess no active temperature control mechanisms and, given sufficient time, their own temperatures equilibrate with those of the surrounding medium: air, soil or water. Temperature influences the rate of chemical reactions and hence impinges upon a plant's physiology. The prevalence of water in plants' anatomy makes freezing temperatures a particular hazard to those species that have not evolved specialized biochemistries and life cycles to cope with this. Air temperature also partially determines transpiration rates, while lethally high temperatures – such as those imposed by fire – may destroy enzymes and plant tissues. In areas where temperatures become lethally high or low or where the correct annual or diurnal temperature cycle does not prevail for a species' ontogeny, it will be unable to survive for an extended period.

Humidity and wind

The primary effects of air humidity and windspeed upon plants are associated with their effects upon transpiration rates. However, high humidities, such as in zones of fog, may enable some plants to acquire a significant proportion of their moisture supply from this source (e.g. Parsons 1960), while the permanently high humidities in some tropical and temperate rainforests permit the existence of an abundant epiphytic vegetation (pl. 4). High and persistent windspeeds may affect plants by abrasion with wind-borne particles, resulting in elimination of ill-adapted species from such sites, and deformation of others persisting there.

Fire

Fire affects plants primarily though its complete or partial destruction of their tissues. However, species differ widely in their susceptibility to such damage. While many species are extremely susceptible even to fires of a low temperature, others possess adaptations such as thick bark, woody tubers and deeply buried rhizome systems that permit survival of severe burns. Indeed, some species even require fire for completion of their life cycle. The serotinous cones of jack pine (*Pinus banksiana*) only open to release seed when exposed to fire temperatures, while the related grasses *Imperata cylindrica* and *I. brasiliensis* of the Old and New World tropics only flower following fire. Adaptations such as these bespeak a prolonged history of fire as an environmental component of higher plant life (pls. 5, 6). Although there once existed considerable controversy over the possibility of naturally set fires, it is now well established that lightning can frequently set fires independently of man (Komarek 1964). This is not to deny that man's activities have produced great changes in the extent and frequency of fires, a topic treated further in chapter 10. Fire, of whatever origin, is today a major modifier of many species' distributions and vegetation types. The distribution of species such as *Imperata cylindrica*, which are at a competitive advantage in regularly fired environments, is closely correlated with recent fires. Other species, susceptible to fire, are often excluded from otherwise hospitable habitats by the frequency of fires there. The existence of many vegetation types, notably many tropical savannas, appears to have been markedly affected by a high frequency of fire (Batchelder and Hirt 1966).

Interactive effects

Few plants grow isolated from other organisms and few abiotic influences are unqualified by the effect of some adjacent or associated organism. These interactive effects with other organisms may be functionally separated into those involving some direct effect through physical contact (symbiosis, parasitism, predation) and those in which the effect is exercised through an intervening environmental medium (competition, allelopathy and other environmental alterations). This distinction is significant to interpretations of interactive effects for, while in the former a 1:1 effect and response may be expected, in the latter, effects may be partially dissipated in the intervening ambient medium, precluding so close a relationship between the two organisms.

It is seldom possible clearly to distinguish between symbiosis and parasitism. Symbiosis is usually defined as the living together of two or more organisms to their mutual benefit, while parasitism requires one organism (the parasite) to be living off another organism (the host plant) to its detriment. However, the subtleties of the interactive effect make it seldom possible to designate any interaction as clearly one or the other and only certain *facets* of most interactions can be specified in this way.

The symbiotic relationships between higher plants and various micro-
-organisms are many and varied. Two of the most important involve those
with nitrogen fixing bacteria in root nodules (pl. 7) and fungal mycorrhiza
on the surfaces of some higher plant roots. Nitrogen fixing root-nodule
bacteria are most widespread in the family Leguminosae but are being
found increasingly in other taxa. The nitrogen fixed in this way is
available to the host plant either directly or after release into the soil as
the ammonium ion (NH_4^+). The role of mycorrhiza is less fully
understood. However, they appear to be important to the mineral
absorption by the roots of some plants. The failure of some pine
plantations on soils not infected with the appropriate mycorrhiza indicate
their importance to the survival and location of some species.

Parasitic relationships may vary from apparent complete parasitism
to that in which dependence upon, and detrimental effect on, the host
plant is slight. There exist a multitude of parasitic micro-organisms, fungi
and insects, and few plants survive uninfected by one or more of these.
Such forms of parasitism are often instrumental in the death of old plants,
although plants at other phases of their life cycle may also be affected.
Plants weakened by environmental stresses such as sulphur dioxide
fumigation may ultimately succumb due to parasitic infection. Some
parasitic higher plants also exist; these possess root structures capable of
extending into the vascular tissue of the host plant to derive photosynthates. The most widespread group of higher plant parasites comprise the
family Loranthaceae, the mistletoes, partial parasites capable of their own
photosynthesis but still drawing some sustenance from the host plant
(pl. 8). Epiphytic plants and ground-rooted climbers have often been
considered not to be parasites. However, the demonstration by Ogawa *et
al.* (1965) that liana infestations in rainforests of Thailand may eventually
kill supporting trees, requires re-examination of this assumption.

Herbivorous animals, ranging from primitive invertebrates to mammals,
are a distinct and important parasitic group whose entire survival depends
upon consumption of higher plant tissues. While herbivore and host may
exist in quasi-equilibrium for long periods, a population explosion among
the herbivores or introduction to the area of some new herbivore may
result in profound modification of the plant populations existing there.
The periodic locust infestations of eastern Africa and the impact of cattle
grazing in the New World are good examples of these processes.

Predation is a specialized form of parasitism whose distinguishing
feature is rapid or immediate death of the affected plant. For this reason,
the predated plant is normally in its seed or seedling stage and the
predator normally an animal. There exists a growing awareness in the
literature that predation of seeds and seedlings may have a considerable
impact, both proximally and ultimately, on plant species' populations,
especially in the tropics where no annual climatic cycle decimates the
predator population (Janzen 1970).

The indirect effects of one higher plant upon another have frequently been ascribed to competition, a phenomenon whose definition has tended to remain vague. However, in recent years, Muller (1969), Whittaker and Feeney (1971) and others have demonstrated the interactive effects, through chemical toxins, of one plant upon another. Such effects have come to be termed allelopathy and Muller (1969) has suggested that they be clearly separated from competition: the former involve additions to the environment of toxic materials, the latter removal of required resources for growth. Both effects are difficult to analyse. Demonstration of competition normally involves careful experimentation with many densities of competing plants, or such procedures as trenching experiments in the field to remove root competition. Similarly, the difficulty of identifying the chemical toxins responsible for allelopathic effects is considerable. Many of these effects have been demonstrated only in a functional way: aqueous extracts of various plant species' tissues have been shown to inhibit germination or growth of its own or other's offspring, the effective toxin remaining unspecified (e.g. McNaughton 1968). These procedures, too, require careful experimentation.

There remain a number of interactive effects involving micro--environmental alterations which are difficult to clearly place within either competition or allelopathy. The creation of a shade-light environment of atypical spectral composition by some species, and the production of a unique soil environment by the atypical, but non-toxic litter of other species are examples of such effects.

The field environment

Some of the principal environmental conditions contributing to higher plant species' distributions have been discussed above. This knowledge has accumulated over a period of time from work on the autecology of a variety of species. The application of this knowledge to the solution of specific plant geographic problems in field situations produces a suite of secondary methodological problems.

The first of these concerns the question of what variables to measure as possibly influential on plant distributions. As it is seldom practicable to measure an extensive array of variables, it is necessary to exercise considerable judgement when formulating hypotheses on the possible role of specific environmental parameters. As detailed autecological information on the one or more species to be treated is seldom available, one is forced in these situations to draw upon two sorts of information: autecological work on other species of related taxa or habitats, and such insights as may arise from preliminary field observations. In drawing upon autecological information from other taxa, one is usually faced with a bias towards economically valuable species, making the results of tenuous utility when applied elsewhere. This results in great emphasis often having to be placed upon preliminary field observations. In most instances, keen field observation, receptive to the subtleties of environmental and plant population variations, is an essential prerequisite of hypothesis formulation, and cannot be replaced by 'shotgunning' with computer-programmed multiple regression analyses. In practice, the technical limitations of environmental measurements are often an overriding constraint on what is finally measured. It is not accidental that soil pH and maximum/minimum temperatures are two of the most widely measured variables!

A second problem facing the field researcher is the question of what facet to measure of the various environmental variables identified. Few plants exist in a spatially uniform habitat: roots and shoots exist in radically different media, while within these, appreciable variations may occur. The lower branches of a tree often exist in a very different shady micro-environment from the topmost branches in full sunlight. Where to measure the environmental variables is thus a fundamental problem. Similarly, while soil chemical status may be relatively constant through time, soil moisture and most atmospheric conditions are varying constantly with time. What facet of this temporal variation to measure is a further problem which can only be solved through enlightened judgement based on extensive reading and field observation.

An even more vexing methodological question concerns the interpretation of plant/environment relationships in the field. This revolves around the phenomenon of collinearity between environmental variables, and synergistic effects of several variables on the same plant. Few environmental variables have the independence normally assumed in most statistical analyses. For example, an increase in insolation normally brings with it, albeit with some time lag, an increase in temperature, a decrease in relative humidity and other atmospheric changes. Because of such collinearity, several variables may correlate well with some plant distribution yet some, or possibly none, may be causal. Collinearity between measured variables is less serious than between measured and unmeasured variables as, in the latter instance, an observed correlation may be entirely spurious, being based on the effect of an unmeasured condition.

It is now appreciated that synergistic effects on plants by differing environmental variables may be common. For example, the application of several nutrient elements in fertilizer may elicit a greater response in plants than the sum of individual responses when these nutrients are applied separately (table 3.1). While synergistic effects that are proportional can be handled by the more sophisticated analyses of multiple correlation and regression, those which operate non-linearly or only beyond certain threshold values will remain difficult to detect.

These problems of collinearity and synergism in the field environment add considerable complication to the concept of simple limiting factors on species' distributions (cf. Allen 1929). They have led some workers (e.g. Major 1961) to propose that a purely functional approach to environmental relationships be adopted. While such an interpretation may be

Table 3.1 Synergistic effects (positive and negative) of nutrient elements on oil palm yields at Bunsu, Ghana. Yields in lbs of fruit bunches per acre per year

Treatment	Yield	Difference
Control (no P or Ca)	9,257	
P	10,233	+976
Ca	9,631	+374
P + Ca	8,997	−260
Control (no K or Mg)	9,559	
K	9,387	−172
Mg	8,548	−1011
K + Mg	10,624	+1065

Source: Hartley (1968)

acceptable to limited problems in a localized context where purely functional predictions suffice, it would seem inimical to the extension of research to new habitats or to the development of plant geographic theory in general.

Despite the desirability of measuring only those conditions postulated to have some direct effect upon the physiology of a plant, practical difficulties frequently limit such an approach, necessitating cruder, but

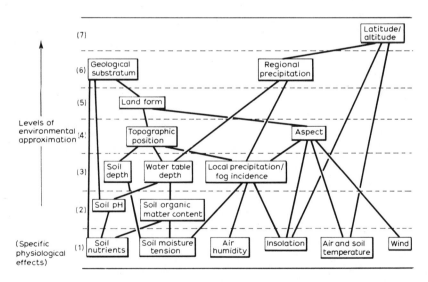

3.4 A hierarchy of field environment approximations. In this schema topographic position and aspect occupy an important place and may be expected to predict, statistically, a high proportion of the variation in local plant distributions.

more easily measured, approximations of these. Environmental conditions, such as insolation, that may impinge directly upon a plant's physiology are often closely correlated with landscape features, such as aspect, that themselves have no direct effect upon plants. However, because these latter features are more easily measured, they provide useful means of approximating environmental conditions in the field and may explain, statistically, a high proportion of plant distributional phenomena (e.g. Eis 1962). Consequently, a hierarchy of environmental approximations may be developed which ranges from those involving specific physiological interactions at the lowest order, to those of a large scale and far removed from such direct physiological interactions, but nonetheless operatively connected. One such hierarchy is presented in fig. 3.4. It is by no means definitive and other more appropriate hierarchies could be developed for specific situations. However, it is essential in the development of these that the 'vertical' linkages in the hierarchy be clearly understood to ensure that legitimate conclusions about plant/environment relationships are drawn.

4 Plant mobility and species' ranges

In chapter 1, the distinction was made between potential ranges of species and their actual ranges. The degree to which a species realizes its potential range, as set by environmental conditions and competitive interactions, is largely a reflection of its ability to migrate: many weedy species approach full potential ranges while less efficient migrators occupy far more limited areas. In contrast, some species of great longevity, notably trees, are able to survive beyond their potential range as relicts until death eliminates them from these sites.

Species' source areas

Plant species begin their range expansion towards potential limits from source areas which may be of two types: evolutionary source areas, and refugia. As discussed previously, the species evolves not as a single plant, but as a population of interbreeding but relatively isolated plants.[1] The area occupied by this population becomes the source area of the species; configuration and extent will be functions of the population's distribution and the frequency of gene interchange. From such evolutionary source areas the species may spread if environmental opportunities are available and competitive restrictions can be overcome. Thus, in the mid-Tertiary, a drying trend in the climate of western North America appears to have led to the progressive spread of drought adapted species from formerly isolated pockets to form a regional desert flora (Axelrod 1958).

In contrast, environmental change (mainly climatic) can result in the restriction of formerly widespread species to one or more isolated refugia. Re-migration from these secondary source areas must take place once environmental conditions ameliorate if the species are once more to

---

[1] The exceptions to this involve polyploidy and hybridization where a new species may be represented, initially, by a single plant.

occupy a wider range. The climatic oscillations and temporary ice advances that occurred throughout the Quaternary have undoubtedly had the most profound impact on higher plant species' ranges since their evolution in the late Cretaceous. Much of the earth's plant cover in higher latitudes reflects the impact of these environmental changes, while recent evidence suggests that the impact within the tropics may have been considerable (Vuilleumier 1971). Some plant species appear not to have re-advanced from refugia as indicated by their present restriction to formerly ice-free areas although the correct interpretation of such restriction is often controversial. In contrast, the evidence from many fossil pollen sequences (chapter 8) indicates that many species have migrated considerable distances in postglacial times. Pollen and macro-fossils from a deposit in northwestern Georgia dated as full glacial form an assemblage whose closest living counterpart is the boreal forest of northern New England, 700 miles distant (Watts 1970). Similarly, species of higher elevations in tropical mountain areas are known to have been present at lower elevations during periods in the Pleistocene (Flenley 1967; Martin 1963; Walker 1970b).

Recent climatic oscillations have likely had an important, although less dramatic, effect in restricting plant species' ranges. In a series of paired photographs from the American southwest encompassing intervals of up to eighty years, Hastings and Turner (1965) have illustrated consistent changes in plant populations which they relate, in part, to climatic change. The more widespread use of fire since early pre-historic times has probably also been a significant restrictor of species' ranges.

The effects of environmental change and 'forced' mobility on plant species populations may be profound. Where rapid migration, or migration across inhospitable terrain, is required, appreciable impoverishment of the flora may occur as the less efficient migrators become extinct. The latter process appears to be the prime reason for the impoverished flora of western Europe, whose species populations were required, during the Pleistocene, to migrate eight times across the mountain barrier formed by the Alps and Pyrenees. In contrast, the flora of eastern North America and East Asia faced fewer barriers to migration and underwent considerably less impoverishment during the same period.

Plant mobility

The life cycles of most plant species contain a mobile phase. Exceptions to this are confined to plants which have lost the facility and reproduce entirely by vegetative propagation. The mobile organ has been given several generic terms such as diaspore or propagule and, in higher plants, comprises the seed. Diaspores are normally small relative to the parent plant, dormant, and at least temporarily viable ensuring self-sufficiency during migration. The rare exceptions are viviparous seeds such as the red mangrove (*Rhizophora mangle*), whose seed germinates on the parent plant prior to being shed.

Plant species have evolved diaspores possessing adaptation to a wide range of dispersal agents, and considerable attention has been devoted to the classification of diaspores on the basis of form or function (Dansereau and Lems 1957; Molinier and Muller 1938; Ridley 1930; Van der Pijl 1972). The commoner modes of dispersal include wind (winged and comose seeds), animals (edible fruit, burred seeds) and water. As this wide range of dispersal mechanisms suggests, species differ widely in the range and pattern of diaspore dispersal. In general, the evolving species has available to it two contrasting strategies for diaspore evolution. The available plant food resources may be channelled into a large number of small diaspores: these will have a wide dispersal, and a high probability of arriving at a habitat suitable for establishment, but will have few food reserves to draw upon during establishment of these. In contrast, the available food may be channelled into a smaller number of large diaspores, ensuring adequate food reserves during establishment but necessitating a more restricted dispersal range and probability of finding a favourable habitat. Salisbury (1942) has demonstrated a tendency for species of 'closed' vegetation in Britain, such as woodlands, to favour the latter strategy, while weedy species of more open habitats favour the former. This appears related to the shady but stable habitat of woodlands contrasting with the sunny, but sporadically available disturbed habitats of ruderal vegetation. In contrast, Baker (1972) has demonstrated a positive correlation between mean seed weight and likelihood of seedling exposure to drought stress, in the California flora.

Movement of plant species over very short ranges may be effectively achieved through vegetative spread. Specialized organs facilitating this process include rhizomes, root suckers and stolons. Although movement over only very short distances is achieved, vegetative propagation provides an extremely reliable migration strategy as the offspring may rely on the parent plant during the establishment period. This permits colonization of otherwise inhospitable habitats. In this way bracken (*Pteridium aquilinum*) may spread into grassland by rhizomes (Watt 1947), vine maple (*Acer circinatum*) may spread through mature coniferous forest in the Pacific Northwest by adventitious rooting from layering stems (Anderson 1969) (pl. 9), and trembling aspen (*Populus tremuloides*) expand by root suckering in subhumid grasslands in Canada (Bird 1961). Also of utility to a species is the facility to sustain itself indefinitely by vegetative propagation in an environment where germination is no longer possible. Fire-weed (*Epilobium angustifolium*) can only germinate and establish on open sites freed of competition (often as a result of fire), yet it is able to survive for prolonged periods during the re-vegetation of the site by suckering from rhizomes in which photosynthates are stored. However, on the negative side, prolonged vegetative propagation may be genetically undesirable, by prohibiting gene interchange within a species population.

Diaspore dormancy

Plant species' diaspores must normally remain dormant for some period of time to ensure dispersal of the organ. Only in a few species does vivipary prevail, whereby germination takes place on the parent plant prior to seed shed. The mechanisms of dormancy are, as yet, poorly understood. However, the available data indicate great variability in the time that species' diaspores may remain dormant and viable. Species also differ in the environmental conditions required to break dormancy: those involved may include light, temperature, moisture, carbon dioxide levels and seed scarification.

The advantages to a species capable of prolonged diaspore dormancy are considerable. A greater time period exists for dispersal and, once dispersed, the diaspore may lie dormant awaiting an environmental change favourable to its germination and establishment. A considerable body of evidence has accumulated which demonstrates that the upper horizons of many soils may contain an appreciable content of buried viable seed (e.g. Olmsted and Curtis, 1947; Major and Pyott 1966; Livingston and Allessio 1968; Roberts 1970; Kellman 1974). In a classic study begun in 1879, Dr Beal and his successors demonstrated experimentally this property of prolonged dormancy of seed while buried (Darlington 1931). Most species possessing this property have been weedy or ruderal types common in disturbed habitats. Such species under pristine conditions normally depend for survival on sporadically available disturbed habitats and such a property is clearly advantageous to their survival.

The phenomenon of buried viable seed is of considerable significance in the dynamics of many types of vegetation (see chapter 5). It provides a reserve flora that ensures that a denuded site, even if somewhat isolated from incoming diaspores, is unlikely to remain bare for long. It may also significantly affect vegetation change. In general, this soil reservoir of buried viable seed has not received the attention that it warrants. Although its elucidation involves lengthy experimentation, it is to be hoped that further research will be forthcoming on its extent and importance.

Short-range dispersal

Even in species capable of disseminating seed for long distances, there exists a tendency for the largest numbers of seed to fall in the immediate vicinity of the producing plant, with a progressive reduction as one moves away from the source (pl. 10). This phenomenon can be observed in the pattern of tree seed dissemination into a clear-cut logged setting (Garman 1951, fig. 4.1). The greatest numbers of seed fall at the periphery of the setting with a progressive decline in density toward the centre of the setting.

Such a simple pattern can be appreciably complicated by peculiarities

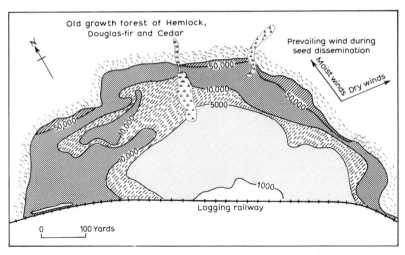

4.1 Distribution of conifer seedlings in a clear-cut logged setting. Seedlings 2—5 years old; densities in seedlings per acre. Source: Garman (1951).

of the dispersing agent. Wind from one prevailing direction may be expected to disperse wind-borne seed predominantly in a single direction. Similarly, preferred lines of movement may be followed by animal dispersers and seed fall peaks occur at nesting sites, seed caches, etc. (Abott and Quink 1970). The result of these processes will be to favour certain directions, distances and patterns of seed dispersal, which may be reflected in the distribution of plant species.

The process of short-range dispersal is also important to the successful reproduction of a species *in situ*. Theoretically, there is a high probability of this taking place as the overwhelming proportion of seed reaching the soil surface beneath a plant will be produced by that plant. However, this may be complicated by several factors. Irregular seed set may result in non-reproduction *in situ* if death occurs in non-seed years. Senescence in the plant may be accompanied by reduced or terminated set of viable seed. The possibility that predator pressure on seeds may, because of easy infestation, suppress seedling establishment in close proximity to parents in tropical forest trees has been raised by Janzen (1970). The further effects of an inhospitable micro-environment induced by the parent tree will be discussed in chapter 5.

Long-range dispersal

Interest in the role of long-range dispersal in establishing plant populations in remote areas has grown out of two anomalous phenomena existing in the earth's plant cover: disjunct species ranges and remote island floras.

Although the distribution of many species populations is sufficiently contiguous that they may be said to form a single range at a continental scale, other species show distinct segregation into one or more populations at such a scale (fig 1.1). Such range disjunctions may be explained in one of two ways: a former more widespread range that has been fragmented by differential species extinction due to environmental change or biotic interactions, or long-range dispersal between areas occupied by the disjunct populations. The first process may explain some disjunctions in areas suffering drastic climatic fluctuations during the Pleistocene. However, in many disjunctions, especially those involving oceanic separations, such an explanation is geologically untenable, leaving long-range dispersal as the sole alternative.

An examination of the role of long-range dispersal is made difficult by the very low frequency of such events and the long time periods that have been available for such events to occur. The establishment of a disjunct population may, for example, be the product of a single exceptionally strong storm dispersing diaspores, available by chance, at any time during the Quaternary. Similarly, the dispersal of plants by salt-water flotation, rafting, or bird migration depends upon comparably exceptional events. Thus, while it is possible to show experimentally that seeds *can* remain viable in dispersal agents for prolonged periods (e.g. Guppy 1917; Olson and Blum 1968; Stephens 1958), it is not possible to recreate the hypothesized event. For this reason, the role and importance of long-range dispersal is likely to remain a debatable issue indefinitely.

The origin of island biotas is largely the product of long-range dispersal and, as such, is a process about which generalization is difficult. However, it has recently been suggested that the *size* of island biotas may be ultimately predictable from such parameters as island size and isolation (MacArthur and Wilson 1967). These authors formulate a model which assumes an island biota in continual change under the influence of continuing immigration of new species and extinction of some existing species. They postulate an equilibrial condition developing after sufficient time, such that the two rates equalize, producing a biota of stable size but continually changing composition. If immigration and extinction rates are proved to be related to factors such as island remoteness and species packing on unit areas, and the postulated model correct, the size of island biotas may prove more predictable than once thought. A limited experimental test of the model on Florida mangrove keys has demonstrated its basic fidelity in that situation (Simberloff and Wilson 1970). However, its reliability in more typical island situations remains unproven.

Although this modelling approach has inadequacies which have been fully specified by Sauer (1969), its appeal lies in its postulate of a dynamic situation, which recent data for vegetation suggest is more correct (chapter 5). Although not applied to the populations of a single species, future developments may make this feasible. The model also holds

Plant mobility and species' ranges · 39

appeal as a practical tool as man's activities continually impose 'island' conditions on biotic reserves and parks, and the long-term conditions of these are seldom postulated. If, for example, the flora of an area is maintained by continual immigration from adjacent areas counteracting local extinctions, then the transformation of such an area into an 'island' may drastically reduce immigration rates and so result in progressive impoverishment of the flora.

The dynamics of species populations

Because plants are of finite life spans, and because physical and biotic environmental conditions are rarely stable for long periods, the possession of dispersable diaspores is an essential, if irregularly utilized, adaptation for survival. This explains the apparently paradoxical phenomenon of abundant 'wasted' seed production by most species. Mature plants of the species continually rain seed about them, almost all of which fails to produce new mature plants. However, the mature plants inevitably die, singly or *en masse,* as a consequence of senesence or environmental castastrophe. At this time, the genotype that has most favourably optimized seed numbers (and hence probability of arriving at a suitable niche), with resources for establishing a new plant once there, will be at a selective advantage. Although innumerable differing evolutionary solutions to this problem have emerged (Salisbury 1942; Baker 1972), in *all* species a relatively large seed output has been preserved as an essential adaptation.

Under temporarily stable environmental conditions, successful seed establishment and maturation within a species' range will tend to equal deaths in the existing mature population. While in many species large numbers of seeds establish initially, severe mortality among seedlings and juveniles normally ensues (fig. 4.2). Under deteriorating environmental conditions, such as those that would be caused by climatic change or the appearance of some new predator, successful reproduction will decline either gradually or precipitously, according to the nature of the change. The survival of the species under these conditions depends upon its ability to disperse diaspores to some new areas possessing habitable niches. The establishment of a viable population in this new area itself depends upon either adequate initial colonization or upon rapid population expansion from a few successful colonists again with the aid of diaspores. In contrast, under ameliorating environmental conditions, the mature plant population may expand its range by means of comparable diaspore dispersal and colonization. The extent to which environmental change can determine the absolute extinction or expansion of a species population depends upon the congruence of this change in space and time. Pleistocene climatic oscillations appear to have involved zonal climatic shifts, resulting in localized species extinctions at one extremity of continental ranges, but expansions at the other extremity. In contrast, the

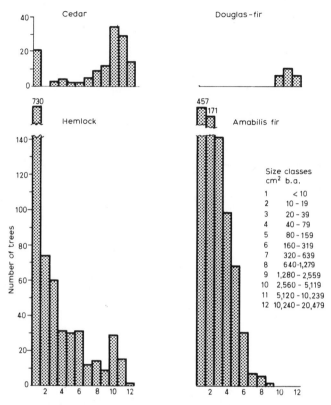

4.2   Tree size-class frequency distributions for the four tree species populations composing a 400-year-old coniferous forest in coastal British Columbia. Douglas-fir relict: cedar severely decimated; hemlock reproducing successfully but irregularly: amabilis fir a successful recent invader. Sample area 1.28 hectares. Source: Shinn (1971).

contemporaneous infestation of the North American chestnut (*Castanea dentata*) by chestnut blight throughout its range appears to have effectively consigned the species to ultimate extinction.

Research on these processes of plant population dynamics is still in its infancy, despite the prolonged study of the phenomenon among animal populations. Moreover, the studies to date (e.g. Putwain *et al.* 1968; Hett 1971; Good and Good 1972) have not yet extended to spatial facets of this phenomenon. Significant future developments in this field of 'geographical' population dynamics may be anticipated.

Part 2/The geography of vegetation

5 Vegetation as an object of study

So far, attention has been focused on the geography of distinct attributes of the earth's plant cover, the individual species. This and two subsequent chapters will treat the plant cover of the earth's surface (or portions of this) as a whole, examining orderly variations in the plant assemblages or vegetation which comprise this. Vegetation may be defined as the total assemblage of plants occupying a given area. It is thus a concept that is spatially defined and is distinct from the individual plant or the plant species which comprise it. Although, theoretically, all classes of plants occupying a given area are part of its vegetation and should be studied as such, in practice it is seldom possible to treat adequately all classes. Consequently, in most studies of vegetation restraints are placed on the definition, limiting it, for example, to all higher plants or to all autotrophic plants. However, this is a constraint required by practicability and does not imply that other groups, such as the soil micro-flora, are unimportant.

Scientific treatment of vegetation presupposes that orderly tendencies exist within it. Geographic treatment pre-supposes that this orderliness is spatial. This requires orderliness in the distribution of plant assemblages.[1] In a highly disordered form, random and non-repetitive assemblages of species may be envisaged occupying given areas, with no consistent inter-species correlations or environmental relationships. The question of the existence and distinctness of species assemblage types (or associations) is central to the whole study of vegetation and has considerable methodological implications treated in chapter 7. It is best approached by considering how plants come together on a bare area to form vegetation.

Vegetation development

Few parts of the earth's surface are entirely devoid of plant cover. Only areas of extreme drought, cold, or continual mechanical disturbance (such

[1] Use of the term 'plant community' has been avoided here because of its conventional (and controversial) connotation of integration between components (see below). Subsequent use of the term has been confined to situations in which such integration is implied.

as beach foreshores) remain in this condition for long. Most other areas become colonized by plants soon after exposure from beneath water bodies, ice, etc., or as deposits of volcanic material, alluvium, glacial debris or aeolian material.

Observation of such sites for prolonged periods of time reveals that the vegetation development on them is often a slow process involving many small changes until a relatively stable and self-perpetuating plant assemblage is achieved. This phenomenon of vegetation development has traditionally been termed 'plant succession'. Two broad categories of plant succession have usually been recognized: primary and secondary succession. Primary successions take place on freshly exposed bare areas, such as river alluvium or glacial till, which have never previously borne a plant cover. Secondary successions develop on sites having previously supported vegetation, which has been partially or wholly destroyed by some catastrophe such as fire, and which may have a considerable impact on the subsequent succession through a remnant flora (Egler 1954) or altered micro-environment (Kellman 1970a) (pl. 11). Many detailed studies of both sorts of succession have been carried out since the classic work of Cowles (1899) on Lake Michigan sand dunes (e.g. Cooper 1928; Billings 1938; Crocker and Major 1955). These studies reveal that plant succession is usually a complex process combining elements of orderliness and predictability with apparent disorderliness introduced by environmental heterogeneity, diaspore availability and chance replacement.

The orderly tendencies in vegetation development were abstracted and formalized by Clements in a series of publications beginning in 1902 and culminating in his *Plant Succession and Indicators* (Clements 1928). Much of the nomenclature still used to describe facets of the phenomenon is attributable to this author. To Clements, who had available only limited empirical data, the earth's plant cover was an orderly and predictable phenomenon which developed along definite pathways to predictable endpoints (see below). Five basic processes in succession were recognized by him: nudation, migration, ecesis, reaction and stabilization. These described initial creation of the bare area (nudation), arrival of available plant diaspores (immigration), establishment of these (ecesis), competition between these established plants and their effects upon their immediate micro-environment (reaction) and ultimate stabilization of the species populations, in an equilibrial condition, at the end phase of the succession.

The type of bare area created prior to the start of succession may have a considerable impact on subsequent developments there. Type of substrate exposed, its degree of consolidation, aeration and chemical status influence the course of primary successions, while the survival of a remnant flora and micro-environment may have an appreciable impact on secondary successions. In primary successions, initial migration of diaspores is often assumed to be a random process, although where propagules are not highly mobile considerable edge effect may prevail (fig. 4.1). Interaction between plants begins when the density or size of colonists leads to mutual interference. This frequently leads to the alteration of the resource environment, notably light levels, leading to the more efficient utilizer eliminating those less capable. However, other 'reactive' effects also operate, especially through plants changing their surrounding soil environment by unique litter types (e.g. Zinke 1962) or chemical byproducts (Del Moral and Cates 1971; Muller 1969; Whittaker and Feeney 1971). A frequent result of these interactive effects is the existence of 'one generation' species populations in successions (pl. 1). Such species become established initially in an environment which changes rapidly as a result of their own reactive effects, or those of associated species, and they are thus unable to establish new offspring to form a second generation there. However, where early colonists are adept vegetative propagators, they may maintain themselves for extended periods by this means (pl. 9). Vegetation change will continue to take place until such time as micro-environmental changes cease, the existing species are capable of reproducing *in situ,* and no further aggressive colonists invade the area. This phase constitutes the theoretical endpoint of the succession.

These processes were thought by Clements to combine in particular areas to produce a limited number of 'seres' or developmental sequences, each named by the initial site condition and all characterized by progressive amelioration of environmental conditions (especially soil moisture supply) to some moderate (or 'mesic') condition. Thus xeric seres were recognized at sites whose initial condition was bare rock with little available moisture, a condition progressively ameliorated by soil development due to rock decomposition and organic matter additions by the developing vegetation. In contrast, hydroseres were recognized at wet sites, such as lake edges and in swamps, where progressive accumulation of debris during vegetation development led to an improvement in soil aeration and an eventual 'mesic' condition. Psamic successions were recognized where moving sand was the main environmental extreme to be overcome, requiring colonists suitably adapted to sand stabilization during the initial stages of succession (pl. 12).

Clements's views on plant succession were far more extensive than implied in this brief summary: in particular, he wedded the concept of succession to that of the 'climax' endpoint (to be discussed below), defining it as 'the universal process of (climax) formation development'. Considerable controversy developed around the concept of plant succession in the early decades of this century, especially between those who, like Clements, regarded succession as solely a progressive 'climax-building' process and others who favoured a more flexible definition. Significant in this literature is Tansley's (1935) distinction between autogenic (self-induced) and allogenic (externally-induced) successions and Cooper's (1926) antithetical definition of succession as 'the universal process of change which is embodied in the great vegetational stream; all vegetation changes, whether internally or externally induced, whether gradual or abrupt, are therefore successional'.

Recent studies of succession, drawing on a less deductive methodology, have treated more fully the non-directional complexities of succession. These involve not only variability in initial site condition, but species availability throughout the succession (including a remnant initial flora in secondary successions), 'external' environmental fluctuations during the course of succession, the results of competition between a wide range of species combinations that may occur during the course of succession, and the element of chance in species replacement. Walker (1970a), using pollen analytical techniques, demonstrates that while certain preferred paths of development have taken place in postglacial hydroseres in Britain, great variability exists in the pathway that may be followed at any one site. Olson (1958) has re-examined the Lake Michigan sand dune succession earlier studied by Cowles, and finds considerable variability in the pathways and endpoints of the succession, occasioned primarily by physiographic variability. Scott (1965) finds comparable complexity existing in the shingle succession at Dungeness. A study of secondary succession on abandoned land in the Philippines (Kellman 1970b) has yielded comparable results. A general successional trend could be detected, but many factors combined to create a great deal of variability in the pathway followed at any site. Certain complicating factors, when considered in combination with time, were found to explain an appreciable proportion of the variability in vegetation composition between sites. Studies of this type have led to a gradual transformation of concepts of succession from those formulated by Clements, to others more similar to those of Cooper (1926). Whittaker's (1953) re-definition of succession effectively summarizes the modern viewpoint:

Succession may thus be thought to occur, not as a series of distinct steps, but as a highly variable and irregular change of populations through time, lacking orderliness or uniformity in detail, though marked by certain uniform over-all tendencies. In its continuity and irregularity and in the sharing of populations in different combinations by different successions, succession is effectively represented by Cooper's (1926) image, after Vestal, of a braided stream.

Future studies of succession may be expected to concentrate more fully on measurable factors that contribute to the process, thereby developing a refined predictive capability for these phenomena.

Vegetation stabilization

Most concepts of vegetation development envisage a progressive diminution of rate of change on any site as this ages, with an endpoint of succession comprising a stable, or relatively stable, assemblage of plants capable or reproducing themselves *in situ* and so maintaining the individual species populations in relatively fixed proportions. Clements coined the term 'climax' for this condition, a term that has received widespread use in the literature, but one that has also generated considerable controversy. Clements restricted the term to describing not simply any stable plant assemblage, but only those that were tightly integrated, assumed to be equilibrated with zonal climate and on sites where no further physiographic or pedologic evolution could produce ongoing changes in vegetation. These he termed climatic climax communities (Clements 1936). These were regarded as the true stable climax assemblages that, in any given area, would inexorably develop through the course of time. When evaluating these somewhat extreme views, one must recall the Davisian concepts of geomorphic cycles accepted at the time of their formulation (Davis 1899). Only under conditions of land surface levelling or peneplanation, the end points of Davis's geomorphic cycles, would one expect cessation of physiographic and pedologic change leaving zonal climate as the main shaper of vegetation. As peneplanation was widely accepted in the environmental sciences in the early twentieth century, it seems probable that Clements drew extensively on this concept when developing his own concepts about the climax community. One must also recall that Marbut (1927), Clements's contemporary in pedology, developed analagous concepts of stable or 'zonal' soil in equilibrium with the zonal climate and climax vegetation. Together these three broad concepts of peneplanation, zonal soils and climax communities appeared to provide a coherent and orderly explanation of processes and phenomena at the earth's surface.

This Clementsian concept of the climatic climax community underwent steady transformation in the first half of the twentieth century as empirical data accumulated in contradiction to it and as Davisian views of peneplanation were rejected. The existence of many different stable vegetation types on differing physiographies in the same climatic zone led to the proposal by Tansley of a 'polyclimax' concept rather than the rigid 'monoclimax' concept of Clements (Tansley 1935). Aubréville (1938), working in supposedly climax tropical rainforest in West Africa, pointed out that many species were not reproducing *in situ* and hence a stable assemblage could not prevail. Watt (1947), drawing on data derived by prolonged observation of permanent plots in bracken and grassland vegetation, pointed to the phenomenon of cyclical fluctuation in this superficially stable assemblage, and suggested that many other vegetation types may exhibit similar processes. Finally, Gleason (1926, 1939), in a now classic paper, enunciated the probabilistic (or stochastic) arguments against the existence of stable climatic climax communities. His 'individualistic' concept of the plant association was based on the observation that both environment and species availability varied in space and time. The result, he concluded, was that no plant assemblage could exist in a stable state but must be continually fluctuating in response to these two groups of factors. Whittaker (1953) reviewed the climax concept extensively, and proposed that it be reinterpreted as a partially stabilized condition in which plant populations, at any given time, were patterned by the existing environmental gradients. He proposed the term 'prevailing climax' to refer to the most abundant vegetation type in existence in some area at a given time.

Today, the climax is no longer regarded as the concrete phenomenon proposed by Clements. Rather, it is interpreted as the floristically relatively stable assemblage of plants which will tend to develop in any area in which species availability is constant and environmental fluctuation slight. As such, its gross morphological features are predictable, but its details vary widely in space and time.

The plant association controversy

Much of the foregoing discussion of vegetation development and stability has implicitly treated the question of the plant association (or plant assemblage type) raised earlier in this chapter. Here, it remains to draw on these and other data raised in earlier chapters, to treat the question of whether the earth's plant cover consists of a mosaic of relatively distinct assemblage types of positively associated species which can form units of study in the geography of vegetation. This has been among the most contentious conceptual issues for students of vegetation in recent decades.

Two opposing schools of thought developed in the twentieth century under the impetus of workers already cited and, with some modification, these viewpoints persist to the present day. At one extreme in the controversy lies the 'organismal' view of the plant community, most strongly enunciated by Clements, and implicitly contained in many classificatory schemes of vegetation. As the name implies, this view equates, in varying degrees, the plant assemblage and a functioning organism. As such, a high degree of integration and species interdependence is envisaged, together with the corollary that species need each other to survive and that the 'organism' is unlikely to survive if constituent 'organs' are missing. At the other extreme lies the 'individualistic' view of the plant association, stated most clearly by Gleason (1926, 1939). This view holds that species interdependence is minimal and assemblages at any site are merely the artifact of present and past environmental opportunities and species availabilities there. Gleason argued that, because these two groups of factors varied constantly, two or more sites were unlikely ever to possess identical species assemblages. Implicit in this argument is the assumption that species' ecological amplitudes are distributed randomly over the range of environmental conditions existing (Goodall 1963).

To evaluate the legitimacy of these two views of vegetation, it is necessary to draw on data about the ecological properties of individual species and a rather limited array of pertinent data from vegetation as a whole.

A number of phenomena may contribute to the tendency for two or more plant species to grow together. Similar (although not necessarily identical) ecological amplitudes are the most obvious of these. Thus, two species whose ecological amplitudes encompass the environmental conditions at some site accessible to both might be expected to grow together there in some form of competitive equilibrium, unless one of the species was so superior a competitor that elimination of the other always ensued. Where sites with similar environmental conditions were found in the landscape, one would expect to find repeatedly the same assemblage of species existing there. Thus in the Pacific Northwest coastal forests wet valley bottoms frequently contain the same limited group of species including devil's club (*Oplopanax horridus*), skunk cabbage (*Lysichiton americanum*) and salmon berry (*Rubus spectabilis*). In contrast, where different environments occupy small areas in close juxtaposition to each other, an apparent association between two or more species may be found, each of which, in reality, occupies different environments. Thus, in the mixed coniferous-deciduous forests of the Great Lakes region, an apparent association of sugar maple (*Acer saccharum*) and eastern hemlock (*Tsuga canadensis*) is attributable to hemlock occupying the wetter sites which are adjacent to the drier interfluves occupied by maple.

A further reason for plant species grouping, and the one most emphasized by the proponents of an organismal view of the plant community, is mutual interdependence between species. Complete mutual interdependence occurs only between such integrated symbionts as lichens (algae and fungi combined) and to a lesser degree in host-parasite relationships. However, although looser, the interdependence between the species which occupies the altered micro-environment surrounding some other species is nonetheless real. Clements and his students placed great emphasis on such relationships, regarding the larger and more abundant plant species as (physiological) dominants which controlled the micro-environment of vegetation and so determined the assemblage of lesser species growing about them. Such interdependence undoubtedly exists between some plants: the logging of a mature forest often results in the death of the forest floor flora requiring the micro-environmental conditions created by the forest canopy. However, the widespread existence of tightly *interspecific* interdependencies of this sort remains to be demonstrated. Were such close interspecific dependencies to exist, well defined species associations would be expected. In contrast, were the interdependence to be less interspecific, far less clearly defined associations would be required. Adequate examination of this phenomenon requires detailed experimentation and, regrettably, as yet no such data exist.

However, two sorts of evidence provide some clue to the problem's solution. Whittaker (1956) has presented data from forest vegetation in the Great Smoky Mountains showing that species distributions along environmental gradients are independent, with no tendency for species to cluster into 'associations' along these (fig. 5.1). Secondly, many fossil pollen assemblages and sequences (e.g. Watts 1971; Wright 1968) show

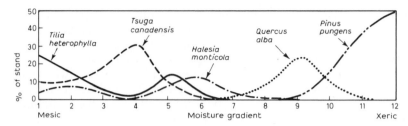

5.1 Abundance of five tree species along a moisture gradient in the Great Smoky Mountains. Note bimodal abundance curves for *Tilia heterophylla* and *Halesia monticola*, suggesting ecotypic divergence within these species or competitive displacement by *Tsuga canadensis*. Source: Whittaker (1956).

changes in species assemblages that are quite incompatible with the concept of integrated communities. These data, together with recent data on the variability of plant successions, strongly suggest that tight integration need not necessarily exist.

Several distruptive processes can be identified as potentially important sources of this observed disorderliness in vegetation. The existence of environmental fluctuations, mainly climatic, which were emphasized by Gleason, have been amply demonstrated by recent climatic records and such palaeo-climatic techniques as dendrochronology (chapter 8). Such fluctuations would continually disrupt the evolution of stable species associations. The self-induced, or autogenic, cycles demonstrated by Watt (1947) and others (e.g. Barrow *et al.* 1968) could have a similar effect although where regularly cyclical, the species association may be envisaged as having a temporal dimension. Finally, the stochastic element which is introduced by the finite life span of constituent plants is perhaps a major source of disorderliness. Upon death of a plant, a vacant micro-habitat or niche becomes available for occupation by another plant. Theoretically, one would expect the new juvenile to be of the same species as the dead plant replaced: the immediate site of a mature plant is normally heavily seeded by this plant during its life cycle. However, were such seedling to decline prior to death of the replaced plant, where a micro-environment inhospitable to offspring establishment was created by this plant, or where other detrimental effects of proximity, such as predator infestation (Janzen 1970) existed, reproduction of the species *in situ* would be improbable. However, even where such perils do not await the offspring of a dying plant, the possibility of non-replacement still exists: chance seeding of the area by the seed of another species at the appropriate instant would achieve this. There thus exists a stochastic element to vegetation composition, imparted by the reproductive biology and diaspore mobility of plant species. Such processes will contribute to disorderliness in vegetation.

The existence of these two groups of counteracting processes in vegetation, one promoting spatial orderliness, the other disorderliness, makes difficult any definitive statements about the plant association controversy. However, recent literature strongly suggests that neither extremity of spatial form is universal in the earth's plant cover. Truly discrete species assemblages have yet to be demonstrated by data gathered in an unbiased manner. Yet few plant assemblages fail to show some degree of repetitiveness, although each is in some way unique or individualistic. Within the broad grey zone between the two extremes lies most vegetation, reflecting both order and disorder in varying degrees. In areas of low species diversity and sharp environmental discontinuities, relatively distinct and repetitive species assemblages may be found with little intergradation between these. In contrast, in areas of great diversity and gradual environmental change, repetitiveness in composition is very difficult to discern and continual intergradation prevails. Consequently, a flexible conceptualization of vegetation seems necessary, encompassing both a partially ordered phenomenon and one whose degree of orderliness may vary with situation. The methodological problems faced when analysing vegetation data will be dealt with in chapter 8. However, it is necessary to first examine how such data are gathered in the field.

6 Vegetation data gathering

Because of the complexity of vegetation, and its continuity over the landscape, a requirement for selectivity pervades all facets of vegetation data gathering. It is quite impractical to consider any total description of vegetation and a complete spatial enumeration is also seldom feasible. Consequently, selectivity in what *sorts* of data to gather and *where* to gather it is of paramount importance.

Vegetation description

Vegetation, defined as the total assemblage of plants occupying a given area, is a phenomenon of great complexity. Because of this, simple descriptions of vegetation, such as a species list or a description of its physiognomy, are of very limited utility. A physiognomic description of vegetation, such as a 'profile diagram' (Davis and Richards 1933–4), or various symbolic notations (e.g. Küchler 1949; Dansereau 1958) does little more than provide some indication of the gross appearance of the vegetation. Apart from some limited roles, such as in military operations, these descriptions have little utility. The more sophisticated life form system of Raunkiaer (1934) is also limited by its emphasis on one selected attribute: perenniating bud position. A simple list of the species comprising the vegetation is similarly limited. It provides a statement of the floristic ingredients present, but says nothing of the mix in which they occur. Perhaps more significantly, neither a species list nor a physiognomic description permit any exploration of the functioning of the vegetation which is usually a central objective of vegetation analysis.

To fulfil this objective, there is a need for descriptions of vegetation to include not only a species list but also some abundance measure of each species present. However, within this overall requirement, the detailed procedures adopted must rest upon the specific objectives of the data gathering exercise and the morphology of the species comprising the vegetation. Objectives will specify the most desirable measures of species abundance. For example, data gathering in forestry normally involves an emphasis upon wood volume, especially that of economically valuable species. Similarly, a study of tree regeneration would emphasize detailed counts of individual trees of all sizes and ages. However, application of these most desirable measures must often be tempered by what is practicable given the morphology of the species involved. For example, despite the desirability of counting the individual plants of some species, the operation becomes quite impractical with many species that are spreading vegetatively. The individual shoots of the rhizomatous bracken fern *Pteridium aquilinum* can be readily counted, yet it is seldom possible to determine individual plants in the field. The great variability in inter-species morphology often requires that a range of abundance measures be adopted even within a single stand of vegetation.

These two constraints on abundance measures of species have led to a wide range of measures being adopted. However, convergences in objectives and plant morphology have led to an emphasis upon only a small number of such measures in the literature, which are briefly discussed here.

A measurement of plant *density*, in terms of individuals of a species per unit area, is one of the most frequently made enumerations. Clearly, this is only feasible where individual plants are distinguishable, as in the case of many trees, although a density count of pseudo-individuals, such as shoots of bracken, may also be made. Constraints may need to be placed upon the sizes of individuals to be counted in this way, there being normally a great abundance of seedlings of most species present, but progressively fewer individuals of greater maturity (fig. 4.2).

A major limitation of density measurements is their pooled nature and, consequently, their failure to give any indication of spatial variability in species abundance. To assess this, a second measure, species *frequency* of occurrence over an area, has often been used. This is measured by the number of sites at which a species is found within the described area. An array of sample plots is normally used in sampling, and the number of plots in which the species is found is expressed as a percentage of the total number used. The measure essentially reflects the degree of dispersion of the species throughout the sampled area and is a valuable adjunct to a density measurement: two species of similar densities may have widely differing frequencies, one being concentrated in a single area, the other being widely dispersed (fig. 6.1). It is also a measurement that is easily and rapidly made, requiring only the recording of species' presence within each plot, and so has been widely used in surveys.

For low-growing vegetation, especially that composed of species spreading vegetatively, a measurement of percent ground *cover* must frequently be resorted to. This is usually based on a visual assessment of the proportion of a small plot covered by the species. However, even when the plot is cross-gridded to aid in this estimate, the procedure must

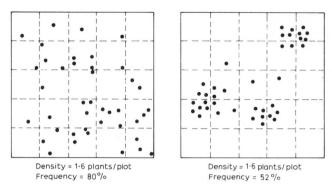

Density = 1·6 plants/plot
Frequency = 80%

Density = 1·6 plants/plot
Frequency = 52%

6.1 Two plant population distributions of similar density but contrasting frequency.

necessarily be rather crude and subjective. The difficulty of applying a comparable measurement to trees has led to the use of measurements made on the accessible lower trunk. The most widely made has been *diameter at breast height* (DBH). The *basal area* at the same height (assuming a circular trunk cross section) is perhaps even more useful because of its additive properties. Using this, the basal area of tree species may be expressed in square meters per hectare of ground area, or in other comparable units.

A further measurement that is of particular use in studies of vegetation productivity (see chapter 9) is the standing crop or *biomass* of the species. This is normally expressed on the basis of dry weight of plant tissue per unit area and, for practical reasons, is usually limited to above-ground parts of the plant.

Some workers have attempted to combine the advantages of several types of measurement by integrating these into a single index. The most widely used of these has been the 'importance index' of Curtis and McIntosh (1951). This utilizes density, frequency and dominance (basal area) measures, each expressed on a percentage scale, which, when summed, yields an index varying from 0 to 300. Although such summed indices have been criticized because of the arbitrariness in assigning ranks to each component, such subjectivity in selection is an inescapable part of *any* abundance measurement.

Finally, there exist a number of qualitatively variable rating scales of abundance, the most commonly used being that of Braun-Blanquet (1932) and its derivatives. Although widely used in the past, such scales have little to recommend them as numbers assigned to ranks cannot legitimately be used in subsequent statistical computations.

Vegetation sampling

The need for vegetation sampling arises out of the impracticability, in most instances, of making a complete enumeration of the vegetation of an area. All sampling attempts a maximization of accurate data about the population treated within certain practical constraints usually set by time and manpower available. The development of a suitable sampling design for any situation must be based both on the requirements imposed by sampling theory and on the immediate objectives of the data gathering exercise.

Sampling almost invariably draws data from a dispersed array of sites within the area treated. This ensures that spatial heterogeneity within the vegetation, which simple inspection often fails to detect, is encompassed in the sample. Most frequently, a sampling plot is used at each site, within which species abundances are counted. Such plots have traditionally been called *quadrats*. These have conventionally been square in shape, but other shapes, notably circles, involve less edge per unit area and, theoretically, are preferable because of the sources of error associated with plot edges. However, it is seldom practicable to locate circular plots in vegetation other than low-growing covers where circular hoops can be used. Instead, rectangles, which also reduce edge effect, are often used. Rectangular plots, placed with long axes parallel to ground slope, have been shown to provide more efficient samples than squares in the deciduous forests of the southeastern United States (Bormann 1953) presumably because of slope-related variability in the vegetation.

Quadrat sizes vary widely. Normally a size is sought that is adequate to ensure that at least several plants can be counted within each quadrat, but not so large as to seriously reduce the number of plots possible to enumerate. However, certain analyses are extremely sensitive to sample plot size (see chapter 7), and where such an analysis is contemplated careful selection in the light of this is required. When the vegetation sample includes a wide range of plant sizes it is necessary to use two or more plot sizes, as it is seldom feasible to count ground vegetation species in a tree-sized quadrat. Where interrelationships between the different vegetation strata are of interest, concentric nested quadrats may be used. A common sample plot size for ground vegetation has been 1m x 1m, and for trees 10m x 10m.

Layout or spatial arrangement of quadrats is one of the more difficult issues in vegetation data gathering. Theoretically, only a randomized arrangement[1] can provide data whose estimate of the population is statistically sound. However, although randomization provides an accurate estimate of the vegetation population taken as a whole, the absence of a thorough dispersion of plots throughout the sampled area may preclude the recognition of spatial patterning within it, and this is frequently a prime objective of the analysis. Where some postulated spatial patterning

[1] In a random arrangement of plots, each plot is located independently of all others. Consequently, as each plot's location is selected, all points within the sampled area have an equal probability of being chosen. Randomization is normally achieved by using pairs of random numbers to locate points from coordinates placed at right angles to each other.

is specified prior to sampling, a stratified random sample may be utilized, with randomization within each stratum. However, vegetation patterns are most often sought inductively from a preliminary survey (chapter 7). For these, some form of regular layout is most appropriate, provided that individual bias in plot location is reduced to a minimum. Such regular layouts are also far more easily achieved. Randomization requires the establishment of orthogonal coordinates and the location of each plot relative to these, while a regular layout may be achieved by compass traversing of the area. Such regular layouts can ensure that data are drawn from all sectors of the area treated and, although bias enters into the selection of inter-plot intervals and grid orientation, the location of individual plots is subsequently free of localized bias occasioned by vegetation or topography. Where gradients in the vegetation, such as those associated with topography, are of prime concern, a regular arrangement may take the form of one or more lines of quadrats orientated along the gradient. If an intense sample is required in such a situation, continuous belt transects may be used. Although providing data that are more useful in most vegetation analyses, regularly spaced sample plots, because of their lack of randomization, make estimates of population conditions less reliable and interpretations based on such samples must be suitably qualified. Ideally, spatial patterns detected with a regular layout should be later confirmed with a spatially stratified random sample.

The establishment of the large number of plots required in a normal sample is a time-consuming exercise and in an attempt to alleviate this a number of plotless sampling techniques have been developed. In ground vegetation, sample belt transects can be reduced to a single line stretched between selected points along which plant coverage can be measured (Wilde 1954). Quadrats can be reduced to a single point from which the distance to adjacent plants can be used to derive density measurements (Cottam and Curtis 1949, 1955, 1956). The use of a suitable instrument enables rapid estimation of tree basal areas and wood volumes from selected sample points (Grosenbaugh 1952).

The selection of a suitable sample size is a further difficult problem of vegetation data gathering and one whose resolution frequently depends upon practical considerations of time and manpower. Since any sample size short of a total count involves some sample error, a further expenditure of effort on extending any sample will provide some added refinement of the estimated population properties. The question of what constitutes an adequate sample thus becomes that of deciding whether information gained per unit sample area declines beyond some threshold sample size, making it progressively less worthwhile to extend sampling further. The index of new information that has been most frequently used has been the number of new species enumerated. This can be plotted as a species/area curve (fig. 6.2) which in most vegetation samples tends to rise sharply early in the sample, falling progressively thereafter. Considerable attention has in the past been devoted to finding a point of inflexion on

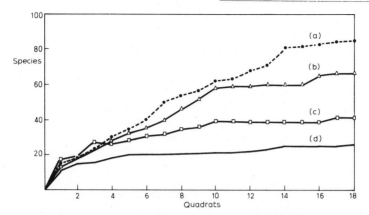

6.2 Selected species/area curves. (a) Primary rainforest. (b) Secondary woodland. (c) Weed vegetation of a corn field. (d) Secondary grassland. 'Steps' in the curves indicate floristic heterogeneity in the vegetation sampled. Source: Kellman (1970b).

this curve thought to be the point at which sampling can cease. However, this search has been based on organismal concepts of the community in which some minimal area for community expression was assumed. Such concepts are no longer tenable and most curves have been shown to be logarithmic (Hopkins 1955) although the ecological basis of this function is unclear. Where the sample extends into some radically new type of vegetation, an abrupt rise and stepped curve results (fig. 6.2). Despite the absence of an inflexion point of such curves, they do indicate a progressive drop in information obtained for a standard increase in sample size. Consequently, they have been widely used as crude indicators of where new information gain tends to 'level off' and sampling may be stopped. If attention is focused on a single species, a mean abundance figure for this calculated at intervals as the sample is enlarged, can be similarly used as a crude indicator of when sampling can cease. At small sample sizes, the figure will fluctuate widely, but become progressively stabilized as the sample mean approaches the mean abundance of the population.

Frequently vegetation data are required for comparisons between stands. For comparability, these data must be drawn from identical sample areas in each stand. This is a particularly difficult sampling problem for seldom is it easy to extend the sample in a stand at some later time, making it necessary to select a suitable sample size at an early stage. This is best achieved by initially taking a very large sample in several stands thought to be representative of the range of vegetation to be met. From species/area curves derived from these early samples, a suitable size can be selected for further work, the excess data in these preliminary stands being rejected from further consideration.

Air photo interpretation of vegetation

The use of aerial photography is a specialized form of vegetation data gathering. Although a complete spatial enumeration of the vegetation is attainable by this technique, the comprehensiveness of the data derivable is not great. For this reason, air photo interpretation of vegetation, while useful for certain specialized purposes, cannot be regarded as a panacea.

Most aerial photography is carried out using vertical camera angles and panchromatic film. The vegetation imagery appearing on such photographs is expressed in four ways: tone, texture, heights, and shapes of larger units. Tones vary from light to dark, textures reflect homogeneity of the canopy and vegetation height is usually determinable using stereoscopic pairs of photographs. This imagery reflects, almost entirely, the canopy vegetation and it is possible to detect gross structural differences in the vegetation from it, such as coniferous forest, broadleaf forest, shrubland, grassland, etc. However, the assemblage of subordinate species remains virtually unknown. Moreoever, where the vegetation is not organized into readily identifiable assemblage types having structural expression, or where spatial variability in the vegetation is essentially continuous, it becomes extremely difficult to identify and locate vegetation types on aerial photographs.

Despite these limitations, air photo interpretation of vegetation is of use for certain specialized purposes, especially where improved forms of imagery, such as colour or infra-red photography, is available. Forest surveyors use aerial photography as a standard tool and wood volumes can be assessed from tree densities and heights. Soil and mineralogical surveys can be greatly speeded if correlations with identifiable vegetation imagery can be established. Infra-red photography has been widely used for early detection of plant disease attacks, especially in crop plants. Aerial photography has also been used in a more comprehensive way during land resource surveys to identify and map the major types of vegetation present, as a preliminary to more intensive examination.

General procedures in air photo interpretation of vegetation involve preliminary designation of the various categories of vegetation present on the photographs. This is followed by field checking at selected points to identify the vegetation reflected in this imagery and possible environmental correlations. Although field checking may yield a wealth of detailed information about subcanopy vegetation, it must be recalled that this cannot legitimately form part of the interpretable data unless a correlation between it and canopy patterns is established.

7 Vegetation analysis

Introduction

The objectives of vegetation analysis may be said to comprise an understanding of the form and function of vegetation, allowing predictions to be made about it both in time and space. Other, more specific objectives can be cited, but these are usually narrowed derivatives of these overall aims. The basic supposition in vegetation analysis is that some form of orderliness or predictability in form and function of the vegetation studied exists. Geographically, the postulated predictability is represented by orderly species assemblages, or associations. However, we have seen that spatial orderliness in vegetation is rarely, if ever, absolute and that species associations are seldom so discrete or widespread that they could form units of study in plant geography. Consequently, the problem of geographical vegetation analysis becomes the detection of those orderly tendencies that do exist in species assemblages, and the relating of these to possibly causal factors which may permit some degree of prediction.

Geographical orderliness in this context is taken to imply an arrangement of plants and species in the landscape in a non-random fashion.[1] Thus, a completely disorderly species distribution is one in which individual plants of the species are randomly distributed, while a disorderly species assemblage is represented by random, non-repetitive, species aggregations. Consequently, vegetation analysis becomes a search for, and measurement of, departures from such spatial randomness in species aggregations.

[1] Greig-Smith (1964) has defined a random distribution as follows: 'In a . . . (random) . . . distribution the probability of finding an individual at a point in the area is the same for all points. Put another way, in a random distribution the presence of one individual does not either raise or lower the probability of another occurring nearby.'

Two modes of approach to vegetation analysis may be distinguished: observational and experimental. Observational approaches, which comprise the vast proportion of studies to date, involve field data gathering for hypothesis testing without any interference with the vegetation, its component species or environment. In contrast, experimental approaches include interference as an essential element to enable control of extraneous variables and synthesis of required conditions for the testing of hypotheses. Although in its infancy, such experimentation is one of the more promising future avenues of research on vegetation.

Observation

The results of field observation of vegetation is usually a tacit or explicit data matrix consisting of different vegetation attributes (species) recorded at a number of different sites (usually quadrats). Species may simply be recorded by presence or absence at a site but are more frequently given some abundance weight where they occur (table 7.1a). The sites where observations have been made may be contiguous, but are more frequently dispersed in some random or regular manner (see chapter 6). This primary vegetation data matrix is usually paralleled by a second matrix, recording the presence and level of various environmental parameters at each site (table 7.1b).

The detection of orderly tendencies in such vegetation data matrices involves two alternate procedures. In the first, orderly tendencies are sought in the vegetation data, and, if detected, are related to environmental conditions via the environmental data matrix or geographical location. The orderliness sought may be relatively homogenous categories (a classificatory approach) or major clines of variation in the vegetation (ordination). An alternative procedure is to stratify the environmental conditions thought to be important to the vegetation into categories or clines, against which the morphology of the vegetation is compared in a search for significant correlations or associations.

Classification

Classifications have been an important element of the scientific approach to many phenomena. They have been used in a variety of contexts: as summaries, pedagogical devices, predictive devices, hypotheses and others. However, all classifications share the common objective of attempting to present a categorization of the phenomenon treated such that the categories are, internally, relatively homogeneous, and, externally, relatively heterogeneous.

In vegetation analysis, classifications have been used to display what are thought to be relatively orderly tendencies, especially in terms of species assemblages. In this role, Major (1958) has made the perceptive comment that 'A classification should be a summary as well as an organization. As a

Table 7.1 Abbreviated vegetation and environment field data matrices from a post-logging secondary stand in coastal British Columbia. Quadrats 1 m × 1 m. Species abundance in individuals per quadrat unless otherwise indicated. Substrate conditions in % ground covered unless otherwise indicated.

(a) *Vegetation data*

Species \ Sites	1	2	3	4	5	6	7	8	9	10	11	12	13
Hypochaeris radicata	9	9	1	3	11	52	26	34	11	6	21	4	7
Pteridium aquilinum (shoots)	17	9	7	6	5	11	7	9	5	7	8	9	11
Epilobium angustifolium (shoots)	3	4	7	2	6	3	2	1	–	2	2	2	1
Thuja plicata	7	1	–	–	2	4	–	1	15	2	6	2	1
Anaphalis margaritaceae (shoots)	1	–	–	1	–	2	3	–	1	4	6	–	–
Trientalis latifolia	–	–	–	4	–	–	–	–	–	4	1	5	3
Polystichum munitum	–	–	–	5	–	1	2	–	–	–	–	2	3
Athyrium filix-femina	–	–	–	1	–	–	–	–	1	1	–	–	–
Dicentra formosa	–	–	–	–	10	12	11	6	–	–	–	–	–
Tsuga heterophylla	–	–	–	–	–	2	–	–	–	–	–	1	–

(b) *Environmental data (substrate types)*

Substrate types \ Sites	1	2	3	4	5	6	7	8	9	10	11	12	13
Mineral soil	5	–	–	–	–	35	–	5	5	10	5	–	–
Rock	–	–	–	–	5	25	–	–	–	–	–	5	–
Burned/unburned	B	B	B	B	U	U	U	B	B	B	B	B	B
litter and humus	30	40	10	30	35	5	10	5	5	5	20	5	5
Rotten wood	–	–	–	–	–	–	10	5	–	5	–	10	–
Frelh slash	–	–	–	–	–	5	–	5	–	–	–	–	–
Charred wood	5	5	75	30	–	–	–	–	25	25	5	5	2
Continuous moss	60	55	15	40	70	30	80	80	65	55	65	80	93

summary it permits advance of a science; as an organization it ossifies it.'
This statement emphasizes the two essential, and contradictory, elements
of a classification. The first is its subjectivity: only two stands of
vegetation which have no species in common can be said to be totally
different and hence classifiable into separate categories with complete
objectivity. Categorization of other stands, having some species in
common, is a subjective process as it incorporates the classifier's
assessment of where category boundaries can best be placed. In this way,
classifications are summaries of the classifier's conclusions about the
vegetation treated. The second element of classification is its organization:
a rigid (even if temporary) framework is imposed upon a phenomenon in
which no absolute categorization may be possible. The dangers that this
property of classification holds for scientific treatment are considerable,
for should a classification become a stable organization, it ceases to draw
upon the external knowledge about the phenomenon that went into its
original formulation, and effectively 'ossifies' scientific treatment of the
subject. The resolution of this conflict is clearly to make classifications
temporary organizations or summaries, reflecting the current state of
knowledge about the phenomenon. These can be replaced by new
summaries as knowledge about the phenomenon develops. Such summary
classifications of vegetation may have a purely scientific function, but
may also be utilitarian: the classification of a forest into different classes
requiring different sorts of management for timber production is an
example of this latter role. Recently, a new role for classifications, that of
hypothesis generation, has been suggested (Lambert and Dale 1964).
Discussion of this will be delayed until modern computer-based classificatory systems are treated.

Returning to the primary vegetation data matrix (table 7.1a), we can
now see that the process of classification involves a categorization of the
sites on the basis of the species which they contain. The simplest form of
classification is one seeking a single level of organization: the placing of
sites into several categories of equal rank. This has been termed a
reticulate classification. However, a more elaborate classification with
several levels is often sought: this has been termed *hierarchical.* Classificatory strategies designed to yield hierarchies may be either *divisive* or
agglomerative. In the former, the data matrix is initially treated as a single
unit and subdivided progressively to yield a hierarchy; the ultimate
subdivision yielding the original individual sites. In the latter, the
individual sites are progressively pooled, on the basis of some criterion of
similarity, yielding a hierarchy by a reverse strategy. The classificatory
strategy may also utilize only a single attribute (species) in each
categorizing decision (a *monothetic* strategy), or some index based on
several attributes (a *polythetic* strategy).

Most early classifications of vegetation were global in scale, using
structure as the prime criterion in a simple divisive strategy. Thus, world
vegetation, taken as a whole, was subdivided into a limited number of

categories such as broadleaf forests, coniferous forests and grasslands. In the early twentieth century, such systems were extended by Tansley (1920), Clements (1928), Beard (1944) and others, to include floristic attributes at lower levels in the subdivision. An initial subdivision into climax *formations* was extended by subdivision of these into *associations,* each possessing internal floristic homogeneity. For example, the deciduous forest formation of eastern North America was subdivided into the following associations: mixed mesophytic, oak-hickory, oak-chestnut, oak-pine, southeastern evergreen, beech-maple, maple-basswood and hemlock-white pine-northern hardwoods (Braun 1950). Although structurally defined vegetation formations continue to be recognized as a useful elementary categorization, expressing a loose environment/life form correlation (see chapter 2), the organismal concept of the formation espoused by Clements is no longer acceptable. Similarly, at lower levels, the great floristic variability of formations in time and space reveal that the associations are not the integrated community groupings they were once thought to be.

In Europe, an alternative classificatory scheme was developed by Braun-Blanquet (1932). Its widespread use to the present requires that it be treated in some detail. The system is an agglomerative hierarchical classification based entirely on floristics: floristic data from sample plots are grouped progressively into higher abstract categories based on floristic homogeneity. Much of our understanding of the system in the English speaking world is attributable to Poore's (1955–6) extensive review of it.

The primary data matrix used is species lists (with a subjective abundance rating for each species) derived from selectively located plots (or 'relevés') thought to lie in floristically homogeneous stands of vegetation. The matrix is inspected, non-homogeneous data rejected, and the species lists intuitively categorized into a number of *abstract* floristic associations using species fidelity as the main criterion. Fidelity is defined as '. . . the more or less rigid limitation of the plant to definite plant communities' (Braun-Blanquet 1932), with five degrees of fidelity recognized. the associations are then grouped into higher hierarchical levels based on the degree of fidelity of the included species. The system has been quite legitimately attacked on several grounds. An initially biased sample is perhaps the most serious error, introducing an initial intuitive bias into the data matrix, which makes subsequent manipulations of this little more than formalization of unspecified field conclusions. Moreover, important parts of the vegetation may remain untreated or be rejected because they are thought to be non-homogeneous. A second major drawback of the system is the unclear specification of the sorting strategy, a procedure that it is claimed needs to be learned by experience. Although a classification, as a summary, should reflect the classifier's conclusions on the morphology of the vegetation, there can be little justification for failure to specify these clearly.

In essence, the system appears to possess most of the undesirable attributes of an organization and few of the desirable features of a summary. It is highly selective of the data that it treats, employs ill-specified strategies and has proven impossible to apply in areas of floristically complex vegetation. Above all, it relegates its user to a role which is little more than that of descriptive technician. Its continued use in vegetation studies appears to reflect more the inertia of the system than its intrinsic value.

Recent classificatory treatments of vegetation have been dominated by the use of mathematical models and high-speed electronic computers. These developments date from the pioneering work of Goodall (1953) on semi-arid 'mallee' scrub vegetation in Australia. Goodall's classificatory strategy is divisive and monothetic, the sites being subdivided on the basis of presence or absence of selected species. Selection of the species was based on their high frequency and highly significant positive association with other species, with the overall aim of obtaining homogeneous groups of sites. The data were progressively subdivided on this basis to yield groups containing all the selected 'critical' species. Sites from which one or more of these species were absent were pooled to yield a residual group which was then treated in a similar fashion. This procedure led ultimately to four homogeneous groups with a residual element appended to one of these with which it shared no significant heterogeneity (fig. 7.1). Goodall mapped these groups and found a reasonable correlation with topography.

A significant further development was the 'association-analysis' technique of Williams and Lambert (1959, 1960), which has become one of the most widely utilized procedures. This, also, is a divisive monothetic system, utilizing selected species' presence or absence as a basis for subdivision. However, it differs from Goodall's technique in being rigorously hierarchical with no pooling of residual sites. The selection of species for subdivision also differs: those showing the greatest associative tendency (either positive or negative) with all other species, being used. Groups found at any selected level in the hierarchy can be utilized for comparison with environmental features, and these authors show several examples of close correlation between the two.

Subsequent developments have witnessed a plethora of alternate techniques utilizing differing mathematical models or modifications of those existent (e.g. Williams and Lambert 1961a, b; MacNaughton-Smith *et al.* 1964; Edwards and Cavalli-Sforza 1965; Williams *et al.* 1966; Orloci 1967; Crawford and Wishart 1968). However, although differing in detail, all have the common property of being specified mathematical models whose properties are fixed prior to classification. This property, which leads to replicability of results from the same data by any user, has been widely cited as providing objectivity in an endeavour that had previously been far too intuitive and subjective. However, the utility of these techniques must be examined in the light of the overall objectives of vegetation classification.

The overall objectives have been cited previously as the detection of

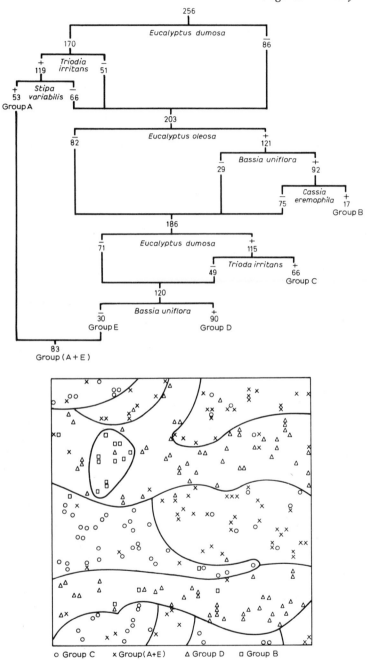

7.1 Goodall's classification of mallee vegetation, and distribution of the four categories derived. Source: Goodall (1953).

orderly categorical tendencies in the vegetation which can then be related to environmental conditions. Furthermore, the classifications should be 'modern' in so far as they reflect all currently available information about the vegetation being treated. Classifications employing fixed mathematical models fulfil the first of these requirements, providing a simplification of data into more comprehensible categories. However, the differing categories that can be derived from the same data using differing models casts serious doubts on whether these are substantially more objective than previous efforts (e.g. Flenley 1969). Clearly objectivity is present in the *application* of the model, yet its original *selection* is entirely subjective. Consequently, while categories may be derived by such techniques which may correlate well with environmental conditions, the inevitable subjective derivation of these categories makes their use as proofs of vegetation/environment correlations impossible, although they may be of predictive value in purely functional terms. The extent to which these classificatory systems fulfil the requirement of modernity is also debatable. Although most are indeed modern in the mathematical sense, few, if any, are up-to-date as to incorporation of current knowledge about the form or function of the vegetation being treated.

Lambert and Dale (1964) have suggested that the real role of these classificatory systems lies in hypothesis generation, rather than in the more traditional one of hypothesis testing. In this way, they suggest that classification of complex vegetation data may reveal patterns which provide new insights into the form or function of vegetation. However, although the use of such classificatory models when treating extremely complex data may provide some starting point for further analysis, in most vegetation data these insights and starting points already exist from field inspection and prior studies. Indeed, they are already implied in the environmental measurements usually made at the time of sampling. Furthermore, the sorts of hypotheses about the vegetation that can be formulated from these models have never been specified — largely, it is felt, because they are not specifiable. Because the orderliness present in vegetation is seldom absolute, it is not possible to demonstrate this in any absolute way. Consequently, as Yarranton (1967) has pointed out, only by invoking the untenable organismal concept of vegetation can one legitimately formulate and test hypotheses about vegetation as a whole. If this is not invoked, one must resort to species-by-species analysis.

It is now necessary to assess the usefulness of the classificatory approach to vegetation analysis in the light of the overall objectives of analysis stated early in this chapter. These comprised an understanding of the form, function and predictability of the vegetation treated. Because vegetation is a multivariate phenomenon for which the organismal concept has been shown to be inappropriate, it is felt that classifications, either intuitive or mathematical, are of little use as modes of *analysis*. Rather, their role appears to lie in *synthesis* of available information about vegetation, reflecting temporary summaries of all that is known about the vegetation to date. In this role, fixed mathematical models are clearly inappropriate, as the strategies must be sufficiently flexible to reflect both the complexity of the vegetation itself and the existing incomplete knowledge about it. This does not imply that unspecified intuitive classifications are acceptable syntheses; only those which clearly specify the modes and reasons for the classificatory decisions can be judged acceptable. The role of mathematical classificatory models appears to be extremely limited. Only in rare instances where the vegetation treated is so unexplored or so complex that no insights exist, does there appear to be a role for these techniques at a very preliminary stage of investigation.

Ordination

Ordination of vegetation data has often been mistakenly regarded as the antithesis of its classification. However, if the vegetation data matrix is visualized as an array with each species providing a separate dimension and sites represented by points fixed according to these (fig. 7.2), it can be seen that the two procedures, although differing, are not antithetical. Each seeks orderly tendencies in the data, but of a different sort. If the data represented is a completely random assemblage of species, an isodiametric array of points would result (Goodall 1963). If, however, as is usually the case, some orderliness exists, this may be represented by the tendency of points to segregate into clusters (fig. 7.2a) or to covary to produce clines (fig. 7.2b). While classification seeks tendencies toward the former, ordination concentrates upon the latter. However, although the data themselves may trend toward one or the other of these sorts of array, it is clear that *either* form of analysis may be applied to *any* data array.

Interest in ordination grew out of the observation that most vegetation data possessed a great deal of continuity in species composition, an

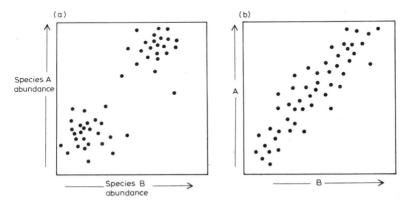

7.2 Positioning of sites in 'vegetation space'. Data for hypothetical, simple two-species stands. (a) Compositional segregation. (b) Compositional continuity along a major cline.

observation that conformed to Gleason's (1926, 1939) individualistic concept of the plant association. Much of the early work on ordination was carried out at the University of Wisconsin by J. T. Curtis and his students. The earliest attempts (Curtis and McIntosh 1951; Brown and Curtis 1952) involved the ordering of sample sites along a single axis. This was achieved by inspection, using the most abundant tree species at each site (the 'leading dominant') such that a relatively continuous transformation from predominance by one species to predominance by others was achieved (fig. 7.3). Plotting of other species' abundances along this axis showed comparable population changes, each species tending to form a curve of bell-shaped form. Certain environmental variables were also shown to correlate partially with this stand ranking. The approach was later extended (Bray and Curtis 1957) by using a floristic similarity index (weighted for species abundance) calculated between all sites to derive a site similarity matrix. From a reciprocal of this matrix ordination axes were extracted geometrically, using the most dissimilar stands as endpoints of an axis along which other stands were positioned relative to their dissimilarity from these two endpoints. In this way it was possible to extract three axes, each orthogonal to the others, representing a crude contraction of the n-dimensional data array. The ranking of sites along these three axes also appeared to be interpretable, the primary axis in this, as in the earlier ordinations, appearing to be related to a successional recovery of the vegetation from past major disturbance. The continuous changes in species populations shown in the displays provided by these

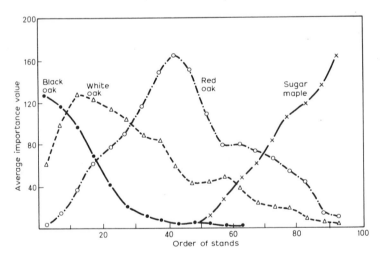

7.3   Changes in abundance of four major tree species along Curtis and McIntosh's earliest ordination of 95 forest stands. Source: Curtis and McIntosh (1951).

techniques were taken by their authors to illustrate the continuous nature of vegetation change in space and the absence of discrete associations. However, some confusion has arisen around this interpretation through a failure to distinguish between the continuity in *composition,* which was demonstrated, and continuity on the *ground.* The latter was not demonstrable from the data which were derived from spatially separated sites. The distinction between the two sorts of continuity are fully discussed by Goodall (1963).

More recent developments in vegetation ordination have, as in the case of classification, drawn upon mathematical models and high speed computers, whose capacity is essential for the formidable computations necessary (e.g. Goodall 1954; Dagnelie 1960; Gower 1966; Orloci 1966; Swan *et al.* 1969; Anderson 1971). Two alternate methodologies have arisen in this work, known respectively as 'principal component analysis' and 'factor analysis' (Lambert and Dale 1964). Principal component analysis aims at a simplified display of sites along a number of axes fewer than those specified by the number of species in the original data matrix. Thus, if the data are markedly orderly, most variability will be associated with a few axes, while the remainder, containing little information, can be ignored. Factor analysis, on the other hand, assumes that fundamental underlying factors exist to which most species will be responding, and which, if species gradients are detectable, may be identified. Individualistic species variation is eliminated from the analysis. However, both methods aim at simplification of the data: they differ in that the factor analysis model selects only common variation among species. Thus while the principal objective of both techniques has been cited as hypothesis generation, the sorts of hypotheses testable are more clearly defined in factor analysis. Both techniques involve elaborate computations which are beyond those not possessing an adequate mathematical background. However, although mathematically more sophisticated than earlier attempts at ordination, these models nonetheless involve a number of subjective decisions. The most important of these is the assumption of linearity in species correlations with influencing factors, and the assumption of an orthogonal arrangement of the axes extracted.

As we have seen, ordination of vegetation data, although differing in detail to classification, is methodologically very similar. Consequently, many of the overall conclusions drawn about the utility of classification for vegetation analysis apply equally to ordination. The main role for the ordinations has been cited as hypothesis generation through the simplification of complex data (Greigh-Smith 1964; Lambert and Dale 1964). As in classification, there appears to be some limited role for ordination in this capacity where the data being treated are sufficiently complex. However, in most instances, hypotheses on the major clines of variation and the principal environmental variables producing these are usually derivable from field observation and prior studies. As Armstrong (1967) has concluded:

Factor analysis may provide a means of evaluating theory or of suggesting revisions in theory. This requires, however, that the theory be explicitly stated prior to the analysis of the data. Otherwise there will be insufficient criteria for the evaluation of the results. If principal components are used for generating hypotheses without an explicit *a priori* analysis, the world will soon be overrun by hypotheses.

However, even this limited role for ordination seems questionable when applied to vegetation for, as Yarranton (1967) points out, it too requires the invocation of an organismal concept of vegetation. These constraints also decrease the value of ordination for predictive purposes although, as in classification, purely functional predictions may be possible about such general phenomena as the environmental parameters that may be expected to elicit the major clines in certain vegetation. One is forced to conclude, then, that the value of ordination as an analytical tool in vegetation studies is in most ways as limited as is classification. Its role in synthesis appears limited to one-dimensional rankings of data along what are thought to be major clines. The extension to loadings of sites along axes and to more than one dimension requires the independent assumption of some geometric system, normally Euclidean. Such synthetic ordinations, representing temporary summaries, should be 'modern' in incorporating all current information on the vegetation with ecological reasons for the ranking being clearly specified. Curtis and McIntosh's (1951) earliest classification appears to come closest to such a synthesis.

Environmental stratification

Although most vegetation analysis has, as its central theme, the search for correlations between the vegetation and environment, this has normally proceeded via a preliminary analysis of the vegetation. However, it is clear that this approach has many pitfalls, associated mainly with the subjectivity inherent in treating a partially ordered multivariate phenomenon. The simpler and more direct approach of seeking, at the outset, correlations between environment and vegetation has not received widespread attention until recently, perhaps due to a lingering organismal concept of vegetation. The works of Whittaker (1956, 1967) and Beschel and Weber (1962) have been the first explicit use of this technique.

The advantages of the approach lie mainly in a reduction of subjectivity. This is clearly present in the selection of environmental parameters on which to stratify, but such a decision is required even in the hypothesis testing of purely vegetational systems, and the further subjectivity of these is avoided. In practice, this approach has yielded much fundamental information about the organization of vegetation and offers considerable scope for spatial and temporal prediction of vegetation from selected environmental variables.

Two alternate approaches to environmental stratification exist: the distinction of environmental gradients and the delineation of environmental categories. The suitability of one or the other will depend upon the field environment being treated, especially its degree of discontinuity, the sorts of vegetation response found, and the overall objectives of the analysis. Environmental gradients have been most widely used under the title 'gradient analysis' (Whittaker 1956, 1967; Whittaker and Niering 1965; Beschel and Weber 1962). This has involved the selection of what are thought to be major, composite, environmental gradients along which the response of the vegetation is examined. In practice, these have usually been resolved into elevation and a complex moisture gradient associated with the degree of exposure. Such complex gradients represent the landscape features discussed in chapter 3 (fig. 3.4). Along these gradients, the abundance of individual species, species richness and structural features of the vegetation may be plotted. A comparable approach has been adopted by Loucks (1962) using three variables (moisture régime, nutrient status and local climate), with an attempt to derive synthetic gradients by means of scalars. The use of categorization as a means of environmental stratification has received little attention so far but clearly offers considerable scope in suitably discontinuous environments such as the terrain of glacial deposition. This approach appears particularly promising for predictive purposes when combined with a probabilistic treatment of species assemblages (see below).

Many vegetational responses to these environmental gradients or categories may be examined, the most fundamental being that of the individual species populations treated by Whittaker (1956) (fig. 5.1). The concentration of certain groups of species, such as those which have been shown to associate positively, can also be related to the environmental stratification and the significance of this relationship tested. In this way, Kellman and Adams (1970) have shown that three distinct weed species associations found in Central American milpas were associated with three distinct physiographies and significantly correlated with soil pH and organic matter content.

The applicability of these analytical techniques based on environmental stratification depend both upon the simplicity of the field environment and the amount of background information available to the investigator. For example, it is difficult to envisage its successful application in tropical rainforest where environmental variations are apparently subtle and their effects upon plants little understood. To this degree the technique is synthetic in that all available knowledge about environmental effects and variations are drawn upon to derive a stratification which is then used for further analysis. In this way, vegetation analysis can progress by procedures of successive approximation common to many scientific disciplines. The utility of the technique will also vary with the scale at which it is applied. Environmental stratification at a global or continental scale can only show variations in gross structure of the vegetation and offer little information on the details of species assemblages.

Less comprehensive analysis of vegetation

A collection of analytical techniques has been developed whose aim is to explore, not vegetation pattern *in toto,* but rather some more limited components of it. Although of more restricted scope, these techniques are of considerable value in clarifying the organization and processes operating in vegetation whose complexity often leads to ambiguous results when subjected to 'whole system' analyses.

Factors controlling the distribution and abundance of individual species within a stand of vegetation can be analysed using multiple correlation and regression techniques, as Blackman and Rutter (1946) have shown for bluebell in British woodland. Yarranton (1967) has argued that, because an individualistic view of the plant association is now widely accepted, vegetation analysis should logically proceed through an accumulation of such single-species analyses. However, the effects of interaction between species would require that, in addition to abiotic environmental variables, all other species in the stand would, in turn, have to be considered independent variables possibly affecting the species under consideration. Apart from the formidable computational task that this would involve, the circularity in specification of dependent/independent variables inherent in the approach appears to limit its utility for analyses of vegetation as a whole.

Considerable attention has been focused on the detailed distribution of individual plants of a species over small areas within stands of vegetation. These analyses have been based on comparisons with the expected distribution of individuals if these were randomly arrayed, which is given by the Poisson series:

$$e^{-m}, me^{-m}, \frac{m^2}{2!e^{-m}}, \frac{m^3}{3!e^{-m}}, \frac{m^4}{4!e^{-m}}, \ldots \ldots$$

where m is the mean number of individuals per sample plot and e the base of natural logarithms (2.7186).[1] Succeeding figures give the probability of quadrats containing 0, 1, 2, 3, 4, etc. individuals, from which the expected number of quadrats in each class can be calculated. The several tests which have been employed in the comparison are discussed by Greig-Smith (1964) and will not be treated here. Departures from a random distribution may be of two sorts: those tending toward an even dispersion, and those tending toward 'clumping'. These two distributions are conventionally called 'regular' and 'contagious' respectively (fig. 7.4).

In practice, most plant species show a contagious distribution at small scales. In such cases, the analysis can be extended by exploring the scale at which contagion (or clumping) is occurring, the location of these clumps,

---

[1] The Poisson expectation applies only where plant numbers are low relative to the total numbers that could grow in the area, and where quadrat sizes are small relative to the total area sampled (Greig-Smith 1964).

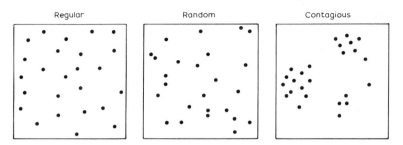

7.4   Random, regular and contagious distributions.

their relationship to each other and to those of other species, and their correlation with environmental conditions. These techniques, which were pioneered by Greig-Smith (1952), are based on repetitive sampling of the same area with increasing quadrat sizes and measuring the variability, as measured by variance, of plant densities between adjacent quadrats at each stage. Where the quadrat size corresponds with the scale of a clump, greatest variability will prevail as those quadrats encompassing a clump will have very high densities while those adjacent will have very few individuals. This disparity will disappear as quadrat size is further increased and each quadrat tends toward containing an equal number of individuals. In practice, the time-consuming procedure of repetitive sampling of the same area with increasing quadrat size has been avoided by sampling initially with a line or block of contiguous small quadrats whose size is progressively doubled by pooling the data from adjacent quadrats. An analysis of variance between adjacent pairs of quadrats at each block size yields a variance figure which tends to rise to a peak corresponding to the scale at which the clump is occurring, thereafter decreasing at larger block sizes (fig. 7.5). The technique is reviewed by Skellam (1952), Thompson (1958), Kershaw (1960), Greig-Smith (1961) and Errington (1973). Kershaw (1961) has also extended the technique to covariance analysis between species, and Yarranton (1969) has recently suggested an allied technique, using regression analysis. Kershaw (1963) distinguishes three forms of pattern: morphological, sociological and environmental. The first refers to the patterning that may appear at small scales in vegetatively spreading species such as vine maple (pl. 9). Environmental pattern represents the effects of a patterned environment being reflected in patterned populations, while sociological pattern reflects patterning produced by species interactions. To these types of pattern may be added a dissemination pattern produced by the concentrations of diaspores in certain locations, notably around parent plants.

These techniques have so far been applied mainly to pattern analysis of individual species. However, the scope that they offer for analysis of the functioning of vegetation is considerable as Shinn (1971) has demonstrated. Unfortunately, the extensive data required for its successful

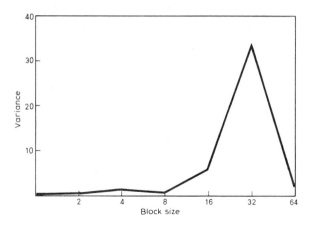

7.5 Pattern analysis of the relict Douglas-fir tree population shown in fig. 4.2. The major scale of pattern at block size 32 represents an area of 80 m x 40 m. Source: Shinn (1971).

application will probably continue to limit its application mainly to individual species distributions in non-arboreal vegetation.

The tendency for plant species to co-occur or covary in abundance within vegetation has also received considerable attention in the literature and a number of measures of this tendency have been developed. The value of the approach lies in its ability to pinpoint orderly spatial tendencies within the vegetation without forcing unconforming or 'generalist' species into a constrained model. In this respect, the approach conforms to the individualistic view of plant associations yet permits insight into their organization. The major limitation of the technique is its dependence on sample quadrat size for, as quadrat sizes are increased, one may expect to find newer and larger numbers of species combinations within them, altering the outcome of the analysis accordingly. Only quadrat sizes carefully selected in the light of plant sizes, ecological relationships, and clearly specified hypotheses can alleviate this short-coming.

Of the two forms of data that may be used in this form of analysis, species presence/absence or species abundance, the former has proved to have both practical and theoretical advantages. Not only are presence/absence data more rapidly gathered, but their use in the analysis avoids the possible complications that species interactions may impose on abundance data (Hurlebert 1969). If two or more species respond positively to the same environmental condition, a positive correlation in their abundance may be found between them at low plant densities. At higher densities, however, competitive effects may appear, leading to a reversal of the correlation trend (fig. 7.6). In such an instance, the measured correlation will not reflect the true nature of the relationship.

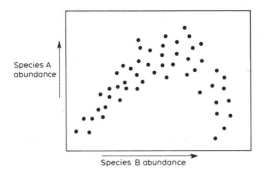

Species A abundance

Species B abundance

7.6   The reversal of a positive correlation trend between the abundance of two species because of competitive interaction at higher densities.

Where presence/absence data are being used, the basic assumption is usually made that the species are independently distributed. If this were so, species *x* should be present in the same proportion of quadrats that contain species *y* in a given sample population, as do *not* contain species *y*. From this it is possible to calculate the *expected* number of joint occurrences of the two species, which can then be compared to the *actual* number. Where these are less than expected, a *negative association* prevails, and where greater than expected, a *positive association* exists. The statistical significance of the associations are usually tested by means of chi-square value.

While a matrix showing significant inter-species associations is revealing in itself, the patterns of positive inter-species association can be displayed visually in a 'constellation' array. This is usually performed by calculating reciprocals of significant positive chi-square values to provide inter-species linkages of a length proportional to their degree of association. These linkages can then be plotted geometrically in two dimensions (e.g. de Vries 1953; Agnew 1961; Kellman and Adams 1970) (fig. 7.7). However, because multidimensional relationships are being compressed into two dimensions, some distortion is often necessary when many inter-species linkages prevail. The resulting array represents the cluster or clusters of species that are tending to associate in the vegetation of the sample area. The location of concentrations of these clusters can be determined from the original raw field data and environmental relationships of these clusters explored by means of correlation analysis and analysis of variance (Kellman and Adams 1970). The arrays are also of use in interpreting floristic organization in vegetation and changes in this during succession (Kellman 1969b). An interesting feature of most analyses, and one that has so far received little attention, is the presence of numbers of positive interspecific associations far in excess of negative associations. While this may be an artifact of sampling designs, it may also reflect a fundamental feature of organization in vegetation whereby positive groupings are not exclusive of other species.

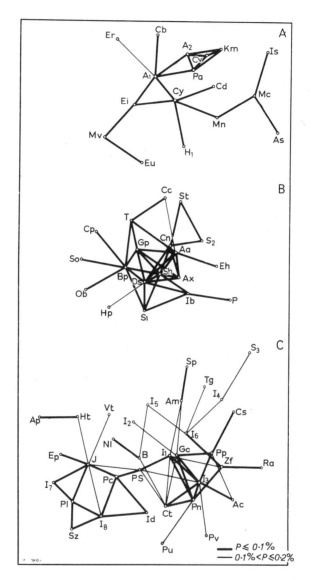

7.7 Positive interspecific associations among weed species in Belizean milpas. (A) Species of permanently used fields on recent valley alluvium. (B) Species of acid soils on old valley terraces. (C) Species of milpas in limestone uplands. Source: Kellman and Adams (1970), simplified.

ABBREVIATIONS

GROUP A

A1	*Acalypha* sp. (402)
A2	*Acalypha* sp. (406)
As	*Amaranthus spinosus*
Cb	*Cenchrus brownii*
Cd	*Commelina diffusa*
Cv	*Cleome viscosa*
Cy	*Cyperus tenuis*
Ei	*Eleusine indica*
Er	*Eleutheranthera ruderalis*
Eu	*Euphorbia hirta*
Hl	*Hybanthus* sp. (420)
Is	*Indigofera subulata*
Km	*Kallstroemia maxima*
Mc·	*Momordica charantia*
Mn	*Melanthera nivea*
Mv	*Mitracarpus villosus*
Pa	*Phyllanthus amarus*

GROUP B

Aa	*Acalypha arvensis*
Ax	*Axonopus compressus*
Bp	*Blechum pyramidatum*
Cc	*Canna coccinea*
Cn	*Chaptalia nutans*
Cp	*Cissampelos pareira*
Ds	*Desmodium scorpiurus*
Eh	*Erechtites hieracifolia*
Gp	*Gonzalagunia panamensis*
Hp	*Hyptis pectinata*
Ib	*Imperata brasiliensis*
Ob	*Oplismenus burmannii*
P	*Phaseolus* sp. (433)
Sl	*Smilax* sp. (559)
S2	*?Securidacea* sp. (637)
Sh	*Spigelia humboldtiana*
So	*Salvia occidentalis*
St	*Solanum torvum*
T	*?Tournefortia* sp. (539)

GROUP C

Ac	*Allophylus comina*
Am	*Aegiphila monstrosa*
Ap	*Aphelandra deppeana*
B	*Bidens* sp. (570)
Cs	*Corchorus siliquosus*
Ct	*Cupania triquetra*
Ep	*Eupatorium pycnocephalum*
Gc	*Guettarda hirta*
Ht	*Hampea trilobata*
I1	*Indeterminate (473)*
I2	*Indeterminate (481)*
I3	*Indeterminate (546)*
I4	*Indeterminate (568)*
I5	*Indeterminate (577)*
I6	*Indeterminate (578)*
I7	*Indeterminate (581)*
I8	*Indeterminate (584)*
Id	*Iresine diffusa*
J	*Justicia* sp. (611)
Mh	*Mimosa hondurana*
Nl	*Neurolaena lobata*
Pc	*Passiflora coriacea*
Pl	*Paspalum langei*
Pn	*Psychotria nervosa*
Pp	*Paspalum* aff. *caespitosum*
Ps	*?Paullinia!=Serjania*
Pu	*Plumeriopsis ahouai*
Pv	*Petrea volubilis*
Ra	*Randia aculeata*
S3	*Sida* sp. (569)
Sp	*Spondias mombin*
Sz	*Stizophyllum perforatum*
Tg	*Tripogandra grandiflora*
Vt	*Vitis tiliifolia*
Zf	*Zamia furfuracea*

Stochastic processes in vegetation

The role of chance in plant distributions was introduced in chapter 1 where the concept of probability was mentioned. While the role of chance is most apparent in the mobile phase of plants' life cycles, it is also present at all times in such processes as climatic fluctuations, disease outbreaks and, indeed, in the uncontrolled variability of unmeasured environmental conditions. Processes in which the outcome depends upon some chance event have been termed stochastic processes and it is obvious that many vegetational processes are of this nature. Despite this, much of the work on vegetation is implicitly mechanistic (cf. Egler 1942b). Although the role of chance is occasionally mentioned, scant attention has been paid to its formal incorporation into the techniques of vegetation analysis. Plant species assemblages in certain environments tend to be treated as absolutely determinable, while successional changes are most often regarded as invariate.

Because so little attention has been devoted to these methodologies, it is only possible to cite some of the avenues along which future developments may take place. At a simple level, and where adequate randomization of sampling has been observed, it is possible to provide the probability of finding a species and its likely local abundance in the sampled area. If ecological research demonstrates the sensitivity of this species to certain environmental conditions, the predictability of its location may be appreciably improved (fig. 7.8). By extending the process, the probability of finding various species assemblages should prove calculable. Alternatively, environmental stratification of the site

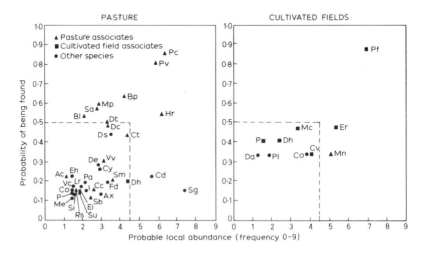

7.8 Probabilities of being found and probable local abundance of the commoner weed species in the upper Belize Valley. For species abbreviations see Kellman (1973)

being treated would permit the most probable species assemblage in each stratum to be determined. The probability of certain temporal changes taking place in vegetation have been treated in a preliminary fashion by Walker (1970a). Anderson (1966) has proposed the use of transition probabilities to explore natural groupings of species, although no application of the technique appears yet to have been attempted.

Although little attention has been devoted to the analysis of stochastic processes in vegetation, this approach appears to offer considerable promise. As older deductive methodologies and deterministic concepts of vegetation are abandoned in favour of viewpoints cognizant of the great complexity of vegetation, it is essential that the role of chance in vegetation phenomena and processes receive formal recognition and treatment.

Experimentation

Traditional approaches to vegetation have been almost wholly observational, despite considerable experimentation on the autecology of individual species. This is hardly surprising as the study of vegetation is a recent science in whose early development extensive observational treatments offered considerable returns at a first approximation level. However, at its present stage of development there are obvious advantages to vegetation science adopting a more experimental approach. Observation has identified many interesting hypotheses and offered solutions at a functional level, whose testing and refinement can only be treated experimentally.

The advantages of experimentation lie in its ability to control, or at least reduce, the influence of extraneous elements that complicate natural vegetation, and in its ability to synthesize specific conditions whose occurrence in the field may be infrequent. However, the approach also has disadvantages, centring around the difficulty of extrapolating the results of laboratory or greenhouse experimentation back to the field. For this reason, field experimentation, involving minimal interference to achieve the desired result, offers greatest promise. However, perhaps the greatest deterrent to experimentation in vegetation has been logistic: almost invariably, lengthy periods of observation are required after the initiation of the experiment, as vegetational processes are seldom rapid.

Four sorts of experimental interference with vegetation may be distinguished: control of the accessibility factor, changes in environmental conditions in the field, dismemberment of plant assemblages and synthesis of plant assemblages.

The vagaries of seed availability are one of the major sources of chance variation in vegetation. Consequently, its control is essential for the accurate identification of environmental and interactive effects. Unfortunately, plant transplantation suffers from the extraneous effects of damage during the process and is impractical for arboreal species.

However, the utility of controlling seed availability has been amply demonstrated by Putwain *et al.* (1968) in their elegant experiments on the population dynamics of *Rumex acetosella.*

Much agronomic research and many range management studies make use of the principle of experimental change in the field environment through such techniques as fertilizer trials and exclosures. However, there has been little specific application of these techniques to elucidate the form and functioning of natural vegetation. A simple, but interesting, experiment was conducted in Malayan rainforest by Symington (1933). Branches of forest trees in a remote area were lopped, permitting entry of full sunlight. The die-back of forest herbs that followed demonstrated the light sensitivity of these species, while the rapid appearance of secondary weedy species strongly suggested their prior existence there in the buried viable seed flora. It is unfortunate that few comparable modern studies have been attempted as significant results may be anticipated.

Much of the contentious debate surrounding the nature of, and integration within, plant communities could long since have been resolved had an experimental approach to community structure been adopted. Despite the obscurity surrounding the role of physiological dominance by certain species in plant assemblages, virtually no experimental treatments of this topic have been attempted. Yet the selective removal of postulated dominant species, and observation of subsequent changes in the assemblage, if any, would largely solve this problem. While Symington's (1933) branch-lopping experiments and studies on root competition partially treat this problem, only recently have Putwain and Harper (1970) applied it more thoroughly to weed assemblages, by selective removal of different species groups with herbicide. Their results, although too detailed to cite here, provided considerable insight into the structure of the weed vegetation being studied. The alternative approach to this problem, through synthesis of artificial assemblages, may prove difficult to achieve save with simple vegetation composed of short-lived species.

Conclusions

Changes in modes of vegetation analysis have reflected, often belatedly, an increased understanding of the vegetation being treated, and changing concepts about its form and functioning. A prevalence, until recently, of observation over experimentation has reflected the youthfulness of the field of study. The search for intrinsic pattern in vegetation has reflected the tacit or explicit belief in an organismal plant community. Recent multivariate analyses of vegetation reflect the same concepts albeit applied with more sophisticated techniques. The increased attention in recent decades to less comprehensive analyses of vegetation has reflected the demise of organismal concepts of the organization of plant assemblages.

However, few new methods of comprehensive analysis of vegetation have yet emerged that are based upon a more individualistic concept of the plant assemblage. Significant future developments in this form of analysis would appear to lie in three directions: (1) a more thorough application of environmental stratification, (2) further analyses of stochastic processes in vegetation, and (3) greater use of experimentation.

Part 3/Other themes in plant geography

8 Terrestrial vegetation history

Introduction

The earth's plant cover as it presently exists has been the focus of attention in the earlier chapters of this book. Although changes of short duration, associated with vegetation development, have been discussed in chapter 5, more prolonged changes, especially those associated with climatic oscillations, have not been treated. The study of such longer term vegetation changes is the topic of this chapter.

Although we may seek no further than man's innate curiosity about the past as a rationale for such historical studies, there exist, also, good theoretical and practical reasons for their pursuance. The adequate explanation of many existing phemomena and processes in the earth's plant cover requires the inclusion of a historical dimension. Had this been available early in this century, it seems probable that our present dynamic interpretation of the earth's plant cover would have evolved far sooner. Moreover, fossil plant assemblages often provide the best available evidence for palaeoenvironmental conditions. Where suitably continuous records of such environmental conditions exist, they provide a useful means of establishing the extent of normal environmental oscillations and the magnitude of recent man-induced change relative to these. The establishment of such recent change indicated by vegetation alteration is in itself a valuable contribution of vegetation history.

The objectives of most studies of vegetation history have so far been twofold: a reconstruction of the past vegetation occupying some site or region, and an interpretation of the environment that existed there. Where suitably continuous records of past vegetation exist, these objectives have been extended to include a reconstruction and interpretation of vege- tation change in the area. These very general objectives have resulted in studies that are largely descriptive and in which the usual procedures of hypothesis formulation and testing have played only a small role. The possibility of more specific objectives for future studies of vegetation history will be discussed later in the chapter.

Historical studies of vegetation require rather different methodologies than treatments of existing vegetation, as well as a group of often elaborate techniques. While the differing methodology can be discussed here, many of the technical details of fossil collection and preparation are beyond the scope of this book. They are adequately treated in texts on palaeobotany and palynology. The time spans encompassed in terrestrial vegetation history can, theoretically, include any period since the appearance of earliest land plants some 420 million years ago. However, for reasons both of practical relevance and available data, most studies have concentrated upon the approximately 10,000 years since the most recent Pleistocene glaciation (table 8.1). During this period, modern plant distributions have become established subsequent to the major disruptions induced by the multiple glaciations of the Pleistocene. However, some attention has also been focused upon the vegetation of the Pleistocene itself, and also that of the foregoing Tertiary period (2–65 million years ago). During this latter period, the true flowering plants (Angiosperms), which first appeared in the Cretaceous, were rapidly displacing the pre-existing Gymnospermous vegetation and diversifying into the many families and genera that compose the great proportion of the earth's existing plant cover.

Table 8.1   Simplified geological table

Era	Period	Epoch	Approximate time span (millions of years before present)
CAINOZOIC	Quaternary	Recent	0.01—present
		Pleistocene	2— 0.01
	Tertiary	Pliocene	11— 2
		Miocene	25— 11
		Oligocene	36— 25
		Eocene	54— 36
		Palaeocene	65— 54
MESOZOIC	Cretaceous		135— 65
	Jurassic		180—135
	Triassic		220—180
PALAEOZOIC	Permian		280—220
	Pennsylvanian		310—280
	Mississippian		355—310
	Devonian		405—355
	Silurian		425—405
	Ordovician		500—425
	Cambrian		600—500
PRECAMBRIAN			Preceding 600

Historical data

Data on past vegetation must be sought in forms that are seldom as complete or easily gathered as data on existing vegetation, and rarely as readily available. Two forms of evidence on past vegetation may be distinguished: fossils and documentary records. Plant fossils of various types and qualities exist in suitable geological deposits. In addition to these, data on vegetation in the more recent past exist in a variety of documentary forms such as written accounts, survey reports, photographic plates and plant collections.

Plant fossils

The term plant fossil may be taken to include all plant remains, or direct evidence of these, that are preserved in geological deposits. Fossils may include virtually unaltered dead plants, or parts of these, the mineralized remains that have replaced these, or the impressions that these organs have left in some geological material. Plant fossils have been classified, somewhat arbitrarily, into micro- and macro-fossils. Micro-fossils include microscopic plant remains either of whole plants (diatoms, algae), distinct organs of these (pollen, spores) or fragments of larger organs (tissue fragment, trichomes, etc.). In practice, most attention has been devoted to fossil pollen and spores. These possess a cell wall, or exine, that is exceptionally resistant to decay, ensuring its prolonged preservation in geological deposits. Pollen and spores are also usually produced in abundance and are readily dispersed to sites of fossilization, thus providing a means of inferring the past regional vegetation surrounding the site of fossilization.

Macro-fossils include all non-microscopic plant remains. Normally, these are composed of detached plant fragments, such as leaves, twigs or seeds, but in some rare instances relatively intact plants (or mineralizations of these) have been found. While macro-fossils have the advantage of generally being more easily identified and related to existing taxa, their poor dissemination often restricts reconstructions of vegetation based on these fossils to that occupying the immediate sites of fossilization.

Fossil preservation requires that the original plant remains be deposited in a site in which organic matter decomposition is eliminated or sharply reduced, physical disintegration precluded, and where ultimate burial of the material will take place. Such sites are most commonly found in shallow freshwater lakes and swamps, protected from wave action and strong currents. Here, subaqueous deposition in an anaerobic environment ensures minimal decomposition of the deposited plant remains, while burial is provided by fine-grained mineral sediments or other organic detritus. Such depositional sites often have considerable temporal continuity, expressed in a vertical sequence of fossils, that allows sequential changes in the surrounding vegetation to be inferred.

Plant remains normally undergo considerable alteration during the fossilization process until fixed in hard geological deposits. Easily decomposed constitutuents are normally rapidly removed leaving the more resistant cell wall structures. In some instances, the entire plant fossil may be removed without collapse of the cavity, leaving a mould which mineral material may eventually fill to form a *cast* of the former plant organ. In other instances, a slow in-filling of individual cells by soluble mineral material such as silica may preserve a near-perfect *petrifaction* of the internal morphology of the plant organ, including its original cell walls. With careful preparation, these can be sectioned to reveal minute internal details, such as annual growth rings in wood. However, most plant fossils undergo compaction, under the stress of accumulating sediments, and are transformed into *compressions* of elemental carbon in which only some details of epidermal morphology are preserved. While such compressions are normally found as thin inter-calated bodies in fine-grained sedimentary rock, occasional great thick-nesses of these occur as coal seams, which have derived sequentially from lignite and peat.

Other sites of fossil preservation are less common but can be locally important. Pollen and spores, whose resistance to decomposition prevails even under sub-aerial conditions, are preserved in soils. The pollen content of soil is a field of active investigation at the present time (Dimbleby 1961). Other fossil assemblages are found at sites where some catastrophic event has provided instantaneous burial of plant remains. The best known of these are cool ash falls that have preserved many partially intact 'petrified forests', especially in western North America. Rapid burial by alluvium, glacial debris, or landslides has also contributed some fossil material. However, these sites, while often preserving a useful instant-aneous record of the vegetation at the time of burial, lack the continuity provided by sites of slow but prolonged sedimentation.

The collection, preparation and preservation of plant fossils is a specialized technical topic that can only be briefly sketched here. The collection of fossils from consolidated sedimentary rocks is largely dependent on adequate geological exposures being available in road cuts, quarries, mines and building sites. However, fossil pollen and spores may be extracted from drill cores of geological deposits. Once collected, an elaborate array of techniques has been developed for the preparation of the material. These include chemical demineralization, thin sectioning, film peeling and electron microscopy.

Recent plant fossils, especially pollen and spores, in unconsolidated deposits are more easily collected using coring devices. Originally coring was confined largely to peat deposits, but recent improvements in coring devices and pollen extraction techniques have allowed more mineralized sediments to be treated. Extraction of pollen from these sediments involves an often elaborate sequence of physical and chemical treatments, including sieving, deflocculation, demineralization and centrifuging, before the residual pollen can be mounted upon a microscope slide for identification and counting.

Documentary data

Reliable documentary evidence of pre-existing vegetation is generally available only for the last few centuries, although more ancient records can occasionally be of use where dramatic changes have taken place (e.g. Mikesell 1969). In general, documentary data on pre-existing vegetation is frequently less reliable than fossil data. However, it is often the only available means of assessing the magnitude of the widespread vegetation changes that have taken place during the last few centuries, changes that are often too rapid to provide a measurable trace in slowly accumulating plant fossil sequences.

Written records of pre-existing vegetation vary widely in quality from very precise accounts to less reliable and often very subjective passing references to vegetation, and such indirect sources as land clearance and timber extraction records. Written records are often limited to brief descriptions of vegetation physiognomy, or to particular plant species such as those of economic significance or bizarre form. Consequently, it is often difficult to develop more than a crude qualitative picture of pre-existing vegetation from these data. However, where vegetation changes have been dramatic, the general trends can be documented from these records, especially when supplemented with detailed field observations of existing vegetation in these areas (e.g. Harris 1965; Sauer 1967). The value of such studies could be augmented appreciably were it possible to supplement the data with pollen records for the more distant past, and experimental treatment of postulated vegetational processes. One of the most reliable documentary records of past vegetation has been the United States Land Survey Reports, in which the species, size and distance to the closest tree in the four quadrants of each section corner were recorded. Several accurate reconstructions of vegetation from these records exist (Kenoyer 1934; Rankin and Davis 1971).

Photographic records of previous vegetation are a special form of documentary data useful in certain situations. Hastings and Turner (1965) have demonstrated their utility in recording changes in desert vegetation in the southwestern United States since 1890, but it is probable that their use in more closed vegetation, or vegetation with fewer characteristic life forms, would be far more circumscribed. The increasing coverage of vertical aerial photography during the last half century will, in the future, provide a valuable means of identifying certain recent changes in vegetation cover.

The advent of widespread botanical collecting during recent centuries has provided a further means of documenting certain aspects of recent vegetation change. However, while providing incontrovertible evidence of species present at the time of collection, plant collections usually involve only an exceedingly small sample of the vegetation, and one that is biased both spatially and taxonomically. Nevertheless, such collections can be used to establish minimal floras of areas at certain periods, and to document recent plant migrations (e.g. Van Steenis 1967).

Historical interpretations

The reconstruction and interpretation of pre-existing vegetation, using data of the sort described above, poses several difficult methodological issues over and above the problems associated with locating adequate data initially. The taxonomic identification of plant fossils and the establish- ment of their relationship to existing taxa is a basic problem associated with using data of this sort. The taxonomy of microfossils is particularly complex, requiring an elaborate morphological nomenclature and com- parison with pollen and spores of living plants. While it is usually possible to identify fossil pollen from the Quaternary as to genus, it is seldom possible to identify species accurately. Also, in some taxa, identification can extend only to families. The taxonomy of more ancient micro-fossils is even less tractable and the pollen content of older geological deposits is used primarily for stratigraphic purposes rather than to infer pre-existing vegetation. The taxonomy of macro-fossils also presents many problems. While Quaternary macro-fossils can usually be assigned to existing species, earlier fossils are often increasingly divergent from modern taxa as a consequence of ongoing evolutionary changes. As a result, new fossil genera and species must often be distinguished. These, however, are based upon scattered and seldom intact specimens, and their taxonomy is frequently controversial.

The development of an adequate chronology for historical data is a further prerequisite of adequate interpretations of vegetation history. While documentary data are normally well dated, plant fossil assemblages are often temporally discontinuous and contained in sediments difficult to date. The advent of radioisotope dating techniques has alleviated the problem considerably, but the high cost of such determinations, limita- tions on their range and frequent absence of datable material continue to present difficulties. For example, postglacial material is normally dated by the so-called carbon-14 technique. However, this technique is unreliable beyond 40,000 years, and other less circumscribed isotope techniques require volcanic material. It is also seldom possible to date more than a few selected points in fossil sedimentary sequences, and the age of intervening sediments must be interpolated.

The identification and dating of a fossil plant assemblage provides only the raw materials for the reconstruction of the past vegetation there. The reconstruction itself is often a rather intuitive procedure for several reasons. Most plant fossil assemblages are a highly selective record of past vegetation because depositional sites are usually wet lowlands, and the assemblage will consequently be dominated by the locally growing swamp species which are often quite atypical of the regional vegetation. While this problem is most severe when only plant macro-fossils are being used, plant pollen and spores, although more widely dispersed, differ appreciably in dispersability. Normally wind pollinated species of large stature are most widely dispersed and so are over-represented in pollen assemblages. Differential fossil production and preservation is a further complication to vegetation reconstructions. Abundant local species may go unrecorded because of poor fossil production or preservation, while abundant pollen producers, such as pine, may dominate a fossil pollen assemblage. While past vegetation reconstructions have been largely intuitive interpretations of the fossil assemblage, in palynology there have been some recent attempts at less arbitrary species weightings (Livingstone 1968) and theoretical treatments of pollen dissemination (Tauber 1967).

The single most important guiding principle in reconstructing and interpreting pre-existing vegetation is the geological concept of uniformitarianism, or the assumption that the present is a key to the past. While this principle has been central to modern geological interpretations, its application to biological phenomena must be accompanied by suitable qualification. Although it is true that the basic physiological processes of terrestrial plant life have remained unchanged through geological time, the processes of organic evolution have nevertheless resulted in an ever-changing array of detailed physiologies, anatomies and ecological properties. Consequently, pre-existing environments, especially those of more distant geological epochs, cannot be easily inferred from fossil plant assemblages. Instead, two more circuitous methodologies must be invoked.

The first of these involves an emphasis upon adaptive morphological features of fossil plants, such as leaf form, irrespective of their taxonomic affinity. While the variability of adaptive morphological features presents logical difficulties to this procedure (chapter 2), the use of a suitably large and taxonomically diverse fossil sample (e.g. Chaney and Sanborn 1933) and the establishment of a modern analogue alleviates these difficulties somewhat. Nevertheless, it is subject to unknown incongruities. A logically more satisfying alternative is to seek analogies between fossil floristic *assemblages* and modern assemblages of identical or closely related taxa. It can be argued, with some justification, that had significant evolutionary changes in ecological properties occurred during the intervening period, parallel change, and consequently continued spatial association, would be unlikely in a diverse and independently evolving flora. In this way, Chaney and Axelrod (1959) have interpreted the environment of the Miocene Mascall flora of Oregon as comparable to that of the lower Mississippi Valley today, because of close floristic similarities between it and the present Mississippi swamp cypress forest.

In palynological studies of Quaternary vegetation change, the search for modern analogues has been central to much of the methodology. In these studies, the pollen content of sedimentary sequences is usually plotted as

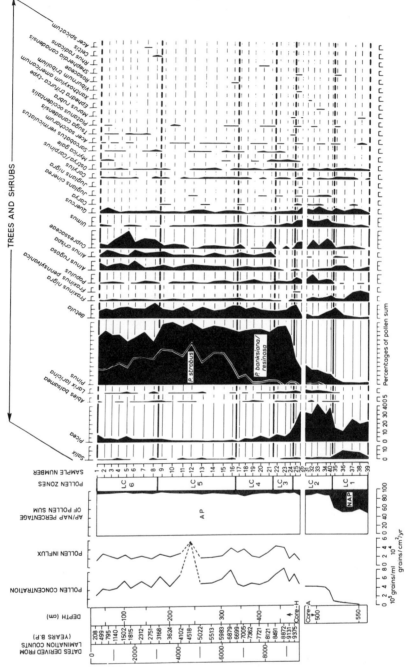

8.1 Portion of a zoned pollen diagram. Source: Craig (1972).

92

a 'pollen diagram', showing the changes (relative or absolute) in the pollen abundance of individual taxa through time (fig. 8.1). Traditionally, these are then 'zoned' into vertical segments thought to be relatively homogenous pollen assemblages, representing a comparatively constant contribution by individual taxa through some period of time. Finally, a modern analogue has been sought for these pollen zones, either in existing vegetation or, more recently, in the pollen rain produced by this. Clearly, this methodology must invoke, at least in part, an integrated view of plant community types that migrate in a relatively intact state over the landscape. In view of our present, more dynamic interpretation of existing vegetation, it is not surprising that incongruities have been found in the pollen record. For example, Davis (1967) finds some late-glacial pollen assemblages in the northern United States that are unlike any of her modern pollen rain types. Furthermore, recent palynological studies in Minnesota (Wright 1968) and the southeastern United States (Watts 1971), have demonstrated a continually fluctuating vegetation assemblage throughout the postglacial, with no clear succession of discrete communities. The possibility of an analogous situation having prevailed in Europe has been raised by West (1963). As our increased understanding of modern vegetation leads to reinterpretations of its form and function, so too must historical interpretations be modified. It seems probable that we can expect considerable modification of past interpretations of vegetation history in the light of these changing concepts. Conversely, detailed studies of vegetation history can illuminate considerably the form and functioning of present-day vegetation.

In sum, the methodological difficulties facing the interpreter of vegetation history bespeak a need for caution in the interpretative process. The absence of data for critical areas or periods, the intricacies of the fossilization process, taxonomic complexity and, above all, a continually evolving interpretation of existing vegetation, require that historical interpretations remain flexible and subject to modification in the light of new data or concepts.

Dendrochronology

Dendrochronology, or tree ring dating, is a rather specialized technique useful in treating some aspects of vegetation history. The technique is based upon the phenomenon that many tree species in temperate areas lay down annual growth rings in their trunks as a result of seasonal formation of xylem cells there. This phenomenon enables one to ascertain the age of the tree itself and, if it is a pioneer species, the dating of the substrate upon which it grows (cf. Everitt 1968). Under suitable conditions, it can also be used to date dead wood, such as that present in archaeological sites, and as a means of interpreting past climatic variation. However, the technique is circumscribed by the availability of trees that exhibit annual rings and, where climatic inferences are being made, trees whose ring morphology (width or density) is suitably responsive to climatic variations. Few trees of the tropics possess annual rings, while the lack of tree ring sensitivity to environmental stimuli in many other areas precludes their use for climatic inferences.

An annual tree ring comprises a sequence from large diameter, thin-walled xylem cells produced in the spring and early summer (early wood) to progressively smaller diameter cells with thicker walls produced in the autumn (late wood). The abrupt change to the next year's early wood marks the ring boundary (fig. 8.2). Xylem cell formation requires the production of photosynthates in the tree canopy and the translocation of these to the sites of cell formation. Consequently, any phenomenon that affects these processes can produce variations in ring morphology. These include environmental conditions that may affect the photosynthetic rate, especially temperature and moisture availability, the canopy size and position, and the remoteness of the site of cell formation from the photosynthetic tissue. Trunk size and age also appear to affect ring widths. Normally, these narrow progressively as a tree ages.

Tree rings may also exhibit lag effects of varying sorts and magnitudes. Current ring formation may trace, not simply to the current year's photosynthesis, but to that of one or more years previously. Similarly,

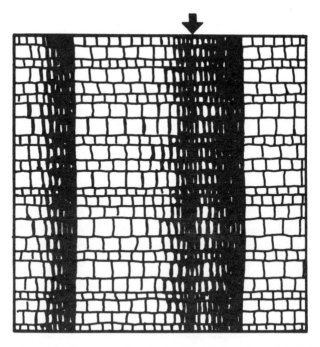

8.2   Cross-section of xylem cells, showing tree ring morphology. A 'false' annual ring also shown (arrow). Source: Stokes and Smiley (1968). Copyright © 1968 M. A. Stokes and T. L. Smiley. Used with permission of McGraw-Hill Book Company.

environmental conditions of a previous winter may be the main determiner of ring width, rather than those of the current growing season. Furthermore, atypical environmental conditions, such as an unseasonal drought, or other disruption of the tree's metabolism, may produce extra 'false' rings or lead to an incomplete or absent annual ring (fig. 8.2). In general, sensitivity of tree rings to climatic conditions has been found to be greatest where trees are growing in open situations under marginal and stressful conditions of temperature (Giddings 1941) or moisture (Fritts *et al.* 1965). Trees in less stressful environments exhibit 'complacent' rings whose widths are both less closely correlated with climate and more subject to the effects of competition from close-growing neighbours.

Dendrochronological techniques have grown considerably in complexity in recent decades and can be sketched only briefly here. Field sampling involves the collection of radial increment cores or entire trunk sections from trees growing in positions that are thought likely to produce tree rings directly sensitive to climatic variability. The cores or sections are mounted on wooden bases and sanded to reveal clear cross-sections of xylem cells. Ring widths are then measured manually with a microscope or automatically using X-ray densiometry of sections. Relative ring width variations are used for cross-correlating individual ring-width sequences and placing each in an absolute chronology. During this process, corrections must be made for false and missing rings.

The derivation of a measure of annual ring width variability involves the establishment of an average trend line for each sequence of measured annual ring widths. In general, ring widths show an early increase as the tree canopy expands, followed by a progressive decrease thereafter as trunk size and canopy remoteness increase. Early attempts at representing this trend involved the calculation of running means for the sequence, then fitting an overall trend line to this by inspection. More recently, mathematical growth functions have been used (Matalas 1962). However, the selection of an appropriate trend line model remains a largely empirical procedure. Absolute ring width is then expressed as a ratio of observed to 'expected' widths to provide a sequence of tree ring 'indices' These indices are normally subjected to tests for significance of inter-tree width correlations at the site, and the records of aberrant trees may be rejected. Finally, average ring width indices for the site are correlated with existing climatic records, and significant correlations used to infer past climatic conditions. These climatic inferences can occasionally be extended beyond the age of existing trees if dead wood of the same species can be found whose ring-width sequence indicates suitable sensitivity and an overlapping life span with existing trees.

Prospect

Much of the work in vegetation history to date has comprised stationary or sequential reconstructions of vegetation in selected areas. Such descriptive treatments have provided a valuable body of data on the composition and change of pre-existing vegetation. However, future historical treatment of vegetation can increasingly address itself to more specific topics. Such enigmatic problems as the origin of atypical vegetation types, the stability through time of species assemblages, and the temporal stability of certain species distributions can only be approached historically. A reinterpretation of many palaeoecological conclusions also seems necessary because of changing concepts on the form and functioning of existing vegetation. The acceptance of a more individualistic view of the plant association will require an increasing emphasis upon indicator species, rather than indicator assemblages, in these interpretations.

The many sophisticated techniques that have been developed for exploring pre-existing vegetation form a valuable resource for the treatment of plant geographic problems. Unfortunately, this very sophistication of technique has resulted in palaeoecological research being concentrated increasingly in a few large laboratories. It is to be hoped that this will not isolate the students of contemporary vegetation from those of past vegetation.

9 Plants and ecosystems: transformations of matter and energy by plants

Introduction

In foregoing chapters, the earth's plant cover has been considered primarily as a dependent phenomenon, influenced by other animate and inanimate phenomena. However, the plant cover invariably has reciprocal effects upon these phenomena. For example, soil conditions may influence vegetation, but this in turn has a reciprocal effect upon the soil by adding organic matter to it and cycling nutrient elements. Similarly, a herbivorous animal population may defoliate a plant population almost to the point of extinction, but then in turn crash as a result of the disappearance of palatable forage. As a result of such reciprocal effects most parts of the earth covered by plants exhibit intricate webs of cause and effect relationships within the biota and between it and the 'physical' environment. While many terms have been used to describe this interacting complex, the one most widely used in the English-speaking world has been the term *ecosystem*. This was coined by Tansley (1935) and defined as 'the whole *system* (in the sense of physics), including not only the organism complex, but also the whole complex of physical factors forming what we call the environment of the biome'.

However, the theoretical ideal of studying such interacting systems as a whole, rather than separate components of these, is difficult to apply for several reasons. Such ecosystems are exceedingly complex, and many important relationships may remain unknown while others, no longer operational, may have artifactual significance. Furthermore, the indefinite boundaries of most ecosystems, their susceptibility to influence by chance intrusions of new genotypes and chance climatic events, make difficult any definition of an area to study as 'an ecosystem'. Interactions between different parts of the system, even when known, are usually qualitatively quite different, making any common quantification of these difficult. For example, in semi-arid environments, potential soil moisture deficits are the principal constraint upon plant biomass. However, the biomass of herbivores feeding upon the vegetation there may be controlled primarily by its protein content, not its total mass. Two such dissimilar relationships are difficult to equate quantitatively in an analysis of the whole interacting complex. Finally, the study of ecosystems requires the invocation of a unique methodology that can encompass reciprocal and reticulate relationships. This is a far more complex methodology than the more usual 'one way' analyses that treat only a dependent phenomenon as potentially influenced by a number of other 'independent' variables (fig. 9.1).

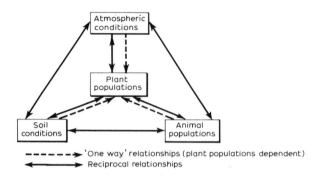

9.1   'One way' and reciprocal relationships between plant populations and other major ecosystem components.

A pioneering attempt at overcoming these difficulties, and the one to which most subsequent ecosystem studies trace, was Lindeman's (1942) proposal that ecosystems be studied by examination of the transfers of energy between components of the trophic, or feeding, web. The use of energy, as a phenomenon common to all components of the system, was seen as overcoming the difficulties posed by qualitative differences in type of interaction. More recent work has added the transfer of organic matter and mineral elements as further common denominators, and virtually all whole-system analyses of ecosystems to date have centred upon exchanges of matter and energy.

Before evaluating this methodology, it is necessary to consider briefly the role that plants play in these exchanges of matter and energy. The special role of plants in the hydrological and geomorphological processes of evapotranspiration, soil water infiltration, surficial runoff and erosion will not be treated here. These are important processes involving the earth's plant cover, but operate as a relatively independent subsystem of the overall ecosystem, and are already adequately treated in the existing geographic literature.

Matter and energy transformations by the individual plant

As outlined in chapter 3, green plants photosynthesize organic compounds from water and carbon dioxide, utilizing incident solar radiation as an energy source. The solar energy is thereafter fixed in chemical form in the photosynthates until released as thermal energy during respiration and the decomposition of organic matter. Total organic matter produced and energy fixed over some unit of time has been termed *gross production*. The proportion of this remaining after respiratory losses in the plant has been termed *net production*, and constitutes growth by the plant. The organic matter composing the plant at any one time has been termed its *biomass* or *standing crop*. Rates of productivity can be measured by periodic harvest of growing plants, or by monitoring the CO_2 exchanges of an enclosed plant. The calorific content of plant and other organic tissues can be determined by oxygen bomb calorimetry. As mentioned in chapter 3, the net energy conversion efficiency of the green plant is quite low, normally less than 1% of the incident solar radiation being fixed.

The net organic matter produced by the plant, and the energy fixed in it, can thereafter follow several pathways. Most directly, the organic matter can be deposited as litter beneath the plant, to be consumed by soil micro-fauna and ultimately decomposed by fungi and bacteria. The rate of dead organic matter decomposition relative to the rate of addition determines the quantity of organic matter stored on or in the soil at any time. Rates of decomposition vary widely, depending both upon environmental conditions and the chemical composition of the dead tissues (Mindeman 1968). An alternate pathway of the plant's net production is its consumption (if palatable) by herbivorous animals that derive their energy supply from the ingested plant material. Some energy is lost in respiration by these consumers, and their body tissues ultimately fall prey to decomposing micro-organisms upon death, or are consumed by a carnivorous predator. Thus, the organic matter produced and energy fixed by the green plant may pass through several trophic levels before ultimate decomposition and energy dissipation.

In addition to the C, H, and O composing most of the organic matter content of plants, other inorganic elements are required for structures and metabolic processes (chapter 3). These are absorbed in an ionic state by plant roots, and translocated to other parts of the plant. In addition to these required elements, others, such as lead, may be absorbed by the plant, but play no active role in its metabolism other than having a toxic effect at high concentrations. The most direct pathway for the return of these elements to the soil is through surficial leaching of the plant tissues by stemflow. However, the more common pathway for return is in plant litter from which the elements are released to the soil during the decomposition process. If plant tissues are consumed, elements may also be channelled through one or more trophic levels before being returned to the soil. Thus elements, like energy, may be stored in plant tissues, animal tissues, dead organic matter, or the surrounding environment. However, a major contrast between the two cycles is the more limited inputs of many elements to the ecosystem. While solar energy is received as a constant, or regularly cyclical, input, many nutrient elements are only slowly released from rock minerals by chemical weathering. Consequently, the evolution by plants of mechanisms for maintaining these elements in relatively closed cycles becomes a major requirement for survival in nutrient-poor soil environments, such as those of the humid tropics.

In their role as transformers of matter and energy, plants are constrained by three factors: environmental resources, other operational environmental conditions and their own genetic potential. The availability of adequate environmental resources for transformation into organic matter is a major constraint. These resources are normally reduced appreciably by competition from adjacent organisms. If resources are limited, the plant may operate less efficiently, or be totally eliminated from the site. A comparable effect may be exercised by other, non-resource, environmental conditions such as temperature. Finally, and perhaps most importantly, the plant's transformations will be determined by its own genetic potential. This will determine both the mode and potential efficiency with which matter and energy are transformed in different environments.

Matter and energy transformations by biotic assemblages

An analysis of the transformations of matter and energy taking place in biotic assemblages is a far more difficult task than the analysis of such transformations by the individual plant. Such assemblages usually comprise a diversity of plant species in various proportions, acting as primary producers, together with a comparably diverse assemblage of herbivores, carnivores and decomposers. Organic matter production, energy fixation and mineral cycling by the vegetation is an integration of these functions by each constituent plant. These functions will, in turn, be affected by the many plant interactions that comprise competition and allelopathy, as well as predation pressure by herbivores and pathogenic organisms.

Under stable biotic conditions, such as would characterize a steady state climax assemblage, environmental resouces of energy, moisture and nutrient elements are shared by the constituent plant species populations existing in competitive equilibrium. This resultant primary production is shared in a comparable way by higher trophic levels. Under such stable conditions, organic matter production and energy fixation will be equalled by decomposition and energy dissipation, while storages in biomass and dead tissues will remain relatively constant through time. Also, mineral element uptake will be equalled by release from dead organic matter and any net loss of elements in runoff or ground water is compensated for by additions from chemical weathering in the solum.

Under unstable biotic conditions, such as those of succession or cyclic oscillations in a climax assemblage, transformations and exchanges of matter and energy change as biotic assemblages change as a result of environmental modification or intrusion of new biota. Under conditions of developmental succession, change normally · involves an increased productive efficiency until some optimum, climax, level is achieved, and an increased storage of matter and energy in live or dead organic matter. However, exceptions often occur as a result of the presence of particular genotypes of atypical form or efficiency. For example, certain pioneer tree species in a tropical secondary succession in Mindanao appeared to enrich preferentially the surface soil beneath them with phosphorus and potassium, while soil organic matter levels beneath the tree-fern phase were temporarily lowered (Kellman 1970b)

Most studies of ecosystems to date have been confined to monitoring the matter and energy transformations of biotic assemblages as a whole, or to measuring the exchanges of matter and energy between trophic levels comprising these assemblages. It is necessary now to evaluate the contribution that this approach has made to our understanding of the complex phenomena that we refer to as ecosystems.

Perspectives on ecosystem analyses to date

Recent decades have witnessed a growing interest in whole-systems analysis in many scientific fields. In ecology, this interest has had its basis both in the ideal of a holistic approach (Evans 1956) and in the practical issues of predicting the productivity of ecosystems or the repercussions upon these of an external stress. There exists a growing body of theory and methods for coping with the analysis of such complex systems that has collectively been termed 'systems analysis' (Wilbanks and Symanski 1968). Theoretically, therefore, the analysis of ecosystems should be a practicable endeavour. However, application of these methods to the analysis of ecosystems has been faced with the fundamental problem of qualitative differences in interactions within the system, a difficulty that does not exist in economic, hydrologic or atmospheric systems. As a means of circumventing this problem, energy and mineral elements have come to be used as common denominators of all ecosystem components, enabling quantification of the relationships between these.

Ecosystem studies of this sort have provided a large body of data describing the energy and matter exchanges within different sorts of ecosystems and the overall organic matter productivity of these (Westlake 1963; Bray and Gorham 1964; Rodin and Brazilevich 1967). These data have been a useful adjunct to the pre-existing information on the physiognomy, biota and environment of these ecosystems. They have also proven of practical value in certain specific problems, such as those of the nutrient budget of shifting cultivation (Nye and Greenland 1960) and the dissemination of pesticides through trophic webs (Woodwell *et al.* 1967).

However, considerable controversy exists as to whether the general approach has provided significant advances in the theoretical or practical analysis of ecosystem functioning. Criticism of the approach centres on a number of interrelated assumptions inherent in its method. These are considered below.

A basic assumption of the approach is that cause and effect relationships in the ecosystem can be meaningfully represented by exchanges of matter and energy. It is certainly true that interactions between organisms and between an organism and its abiotic environment involves exchanges of matter and energy (Spomer 1973). However, the exchanges involve matter and energy in a wide variety of chemical forms, and it is these specialized forms that invariably control cause and effect relationships. For example, the litter fall of two shrubs may be comparable in calorific content and elemental composition; yet if the decomposition of one litter type releases compounds that are strongly allelopathic, its effect upon surrounding plants will be radically different from the non-allelopathic litter. Similarly, an equivalent calorific transfer may be involved in the consumption by some herbivore of plant tissues with or without a toxic cyanide compound. However, the effects of the two transfers upon the herbivore will be radically different. In neither example would the critical chemical control mechanisms be revealed by monitoring only the energy and element transfers. The existence of such transfers would indicate that some sort of interaction existed, but would specify neither the nature nor magnitude of this. Consequently, predictions based only upon a model of such transfers could be quite inaccurate if changes in critical control mechanisms or components took place. The advantages of interchangeability, which makes the approach particularly attractive to physical geographers, should not blind one to this shortcoming. Gross exchanges of matter and energy may be a fundamental measure of cause and effect relationships in geophysical processes, but they are not necessarily so in the diverse biological world.

A further assumption of the method, closely related to the one discussed above, is that organisms are relatively interchangeable within the ecosystem. Thus, the gross matter and energy cycling of the system in some environments is assumed to prevail irrespective of changes that may take place in its biotic components. There are, as yet, insufficient data with which to test this assumption. However, in the light of existing data that suggest a uniqueness of property and function for each species, the proposition would seem most unlikely to be proven correct. For example, the productivity of wood in tropical rainforest environments that was thought to have achieved a ceiling under conditions of natural forest growth (Dawkins 1959) has been dramatically increased by the introduction of exotic genotypes of pine and eucalyptus (Baur 1968). Consequently, it is difficult to conceive of changes in rates of matter and energy transformation, and in the storage of these in different compartments, *not* accompanying the flux of species populations that characterize even climax vegetation (Whittaker 1953). Thus, each slight change in species composition and abundance through time must bring with it comparable changes in matter and energy transformations and storages. Only if species were capable of an inconceivable degree of phenotypic and functional plasticity could the concept of organism interchangeability be accepted as probably correct.

A third assumption of the method, and one closely related to that discussed above, is that ecosystems are fairly rigid (or 'structured') systems, at least in their energetics and trophic relationships. In opposition to this assumption stands the accumulated evidence indicating considerable fluidity in species populations through time, at any site (chapter 5). The possibility of some more basic structure persisting despite such fluidity in populations has already been discussed in the foregoing paragraph. However, a further implication of this assumption is that ecosystems must comprise discrete structured entities that can be identified, delineated on the ground, and studied. Lindeman (1942) was clearly conscious of this implication when he arbitrarily defined an ecosystem as 'the system composed of physical-chemical-biological processes active within a space-time unit of any magnitude'. However, neither Lindeman, nor his successors, has been able to apply the concept successfully except in the atypical situations where sharp environmental discontinuities, such as a land/water boundary, serve to delineate the ecosystem. But such clear delineation seldom exists in exclusively terrestrial situations.

The practical justification for an energy and matter cycling approach to the study of ecosystems is the assumption that data on energy fixation and transformation, and on mineral cycling, are in some way useful to man's welfare. This has been the rationale for much of the data gathered on the earth's major biome types under the auspices of the International Biological Program. However, man is unable to use most of the organic matter productivity of natural ecosystems for either food or as a useful industrial raw material. Nor is the net productivity of these necessarily indicative of their absolute potential (see above). Rather, man requires many specialized sorts of organic products such as digestible carbohydrates, plant and animal proteins and fibres. The productivity of these specialized materials by natural ecosystems usually bears little relation to their total productivity. Consequently, it is the production of *usable* materials by specialized genotypes in synthetic assemblages (i.e. agriculture) that will remain the central problem of human well-being. Undoubtedly, the ecological sciences have a valuable contribution to make to agricultural problems, but this lies primarily in the sphere of assessing the stability of these agricultural systems and recommending measures for ensuring such stability. Measurements of the total productivity of wild ecosystems are of little practical value in such an endeavour.

If these criticisms of the existing approach to ecosystem analysis are accepted, one is forced to conclude that the fruitful results foreseen by Lindeman and others have not yet emerged. While a holistic, quantitative approach has been achieved, this has so far extended only to a description of the gross energetics and element cycling facade of the systems studied, and has failed to probe their intricacies. This conclusion does not belittle the contributions made by this approach to special problems such as nutrient conservation and the dissemination of pollutants, nor does it question the legitimacy of systems analysis *per se*. However, it does question the value of continued application of the existing approach as a means of ecosystem analysis.

Prospect

Despite the deficiencies in ecosystem analysis to date, the achievement of an adequate whole-system analysis of these remains both a theoretical ideal and an endeavour of practical importance. Some possible alternatives are briefly considered below.

A reorientation of methods toward a building-block approach based upon the individual organism would seem worthy of exploration (cf. Gates 1968). There can be little justification for *not* adopting this methodology, until ecosystems can be shown to be relatively independent of their separate components. Furthermore, the reaction of each organism with other components of the system needs to encompass the many subtle forms of interaction that exist over and above simple exchanges of energy and individual elements. Admittedly, it is difficult to conceive of the successful application of these proposals in anything but the most simple system, but perhaps the identification of critical environmental resources and chemical compounds, with a subsequent emphasis upon the quantification of relationships involving these, may reduce the complexity of the task.

A logical intermediate step in such a building-block approach to ecosystem analysis would be the treatment of species population dynamics. Many ecosystem processes (both proximate and ultimate) depend not simply upon biomass, but also upon numbers (i.e. the degree of fragmentation of the biomass). Studies of plant population dynamics have to date concentrated primarily upon density-dependent phenomena and self-regulation. Their successful incorporation into ecosystem analysis would require that attention also be given to inter-population competition, partitioning of environmental resources between populations, predation and allelopathy. A more realistic ecosystem analysis would also require that adequate provision be made for the fluidity that appears to characterize most ecosystems. The identification of the structured, predictable facets of ecosystems, and the incorporation of adequate stochastic functions for those unpredictable facets, would seem essential for a realistic simulation of these.

These proposals suggest an analysis of such great complexity that

attempts at whole-systems analysis may ultimately be precluded. However, if ecosystems are, indeed, far less structured than some have
envisaged, then the very ideal of whole-systems analysis may have to be
questioned. A more fruitful conception of the functional relationships
existing between components of ecosystems may be that of a series of
'waves' generated by each component, whose magnitudes decrease with
increasing distance (spatial or functional) from the component of origin,
as a result of dampening or refraction by other components upon which
they impinge. Under such a schema, the system as a whole can be
conceived of as an array of components occupying a part of the landscape,
whose relative functional and spatial positions at any instant are
determined by the combination of direct and refracted wave patterns to
which each component is subjected. As the entire assemblage is subjected
to continual intrusions or losses of components and waves, and as each
component is itself subject to chance events, the resultant system must
also be subject to unpredictable oscillations. If such a concept of
ecosystem functioning is reasonably accurate, then analysis of these
requires centring upon individual 'wave-generating' components in
sequence, rather than upon an entire system of 'co-equal' components.
Such an approach would preserve the fundamental ecosystem concept of
reciprocity, yet would avoid a resort to excessively simple common
denominators as a means of quantifying the entire system.

10 Man's impact on the plant landscape

Looking about us today, few would question that man's interference with the plant landscape is an all-pervasive and accelerating process. However, the *nature* of this impact has been primarily that of extending and modifying pre-existing processes operating in the earth's plant cover, rather than the creation of totally new processes. Only in its extent and the rapidity of its impact is it unique. Consequently, an adequate understanding of man's role in changing the earth's plant cover must be based upon a knowledge of the form and function of this cover in its pristine condition. For this reason, a consideration of man's plant geographic role has been delayed until the closing pages of this book.

Man's impact on the plant landscape has involved three processes: creation of new environments, dissemination of plants to new areas and channelling of plant evolution. Each process has proceeded sometimes by intent but often accidentally. Together, the three processes have resulted in the modification of species and species assemblages throughout the world, or their destruction and replacement with ruderal and artificial assemblages.

Man's tenure of the earth

Man's impact on the plant landscape may be crudely categorized into three phases: a pre-agricultural phase, an agricultural phase and, most recently, an advanced industrial phase. However, these phases have not always been contemporaneous over the earth, nor have they always progressed in an orderly sequence in all parts of it.

The pre-agricultural phase, during which sustenance was achieved by animal hunting and plant gathering in natural ecosystems, began several million years ago with the appearance of early hominids and persists to the present as the dominant economy in a few isolated areas. During this phase man existed essentially as a subdominant organism, fitting into the trophic web of natural ecosystems as an omnivore. His direct impact upon the plant landscape during this phase was probably slight. Although some plant species may have been gathered to extinction, this seems unlikely in view of their usually diffuse population distributions and the resilience of many plants through vegetative propogation. However, the presence of disjunct populations of edible fruit suggests that some plants may have been accidentally dispersed by pre-agricultural man. A small disjunct population of Garry oak (*Quercus garryana*) exists in the southern Faser Canyon near Yale, British Columbia, separated from the main population of this species by over 100 miles. It seems probable that acorns of the species were brought to this area by Indian salmon fishermen. Similarly, small isolated populations of the tropical American rainforest tree, *Mora excelsa*, in northern Trinidad are thought by Beard (1945) to have been introduced by Indian hunters and gatherers. Some ruderal camp-following plant species may have attached themselves to man during this process as most natural vegetation was not sufficiently disturbed to favour these.

In contrast, man's indirect impact on the earth's plant cover during this phase was probably more profound. The widespread extinction of large herbivorous mammals such as mammoths, mastodons and ground sloths during the Late Pleistocene has been attributed by some authors to over-hunting by prehistoric man (see Martin and Wright 1967). If correct, it seems probable that the disappearance of these large herbivores could have had an appreciable impact upon the vegetation on which they formerly subsisted. However, the most important indirect effect of pre-agricultural man upon the plant landscape is likely to have been through his use of fire for hunting, range improvement or even recreation, and through accidental release. Although considerable debate has surrounded the extent of this process, few can question that some effect was felt once man learned to propagate, and later produce, fire. The plant geographic role of man-induced fire will be returned to below.

In contrast to this ephemeral impact upon the plant landscape by hunting and gathering societies, the advent of agricultural man, possessing the technology to select, propagate and care for chosen plants, has had a profound impact upon the earth's plant cover. The locations, dates and processes involved in the evolution of agriculture remain subjects of ongoing research among archaeologists, geneticists and geographers. However, a comprehensive suite of hypotheses on this subject was presented by C. O. Sauer (1952) who drew upon the work of the Russian geneticist Vavilov (1951) on centres of cultigen diversity and assumed origin, and extensive deductions about the domestication process. A single origin of the agricultural *idea* was postulated with a primary centre (or 'hearth') of plant domestication and agricultural evolution in the mountainous area of South Asia, and subsidiary centres about the world as the innovation spread.

More recent data requires some modification of these postulates, and the overall picture remains obscure. While the existence of a number of centres of cultigen diversity has been confirmed, not all are necessarily centres of primary domestication. Some may be centres of hybridization among crops and between these and their weed associates (Harlan 1961). The oldest dated remains of cultivated plants have come from Turkey and Mexico, both dated at approximately 7000 BC (Helbaek 1964; MacNeish 1965). In the former area, primitive forms of wheat, barley, peas and vetch were being cultivated at the time, and in the latter, pumpkin, squash, annual pepper, amaranth, avocado, cotton and gourds. This suggests an independent and possible contemporaneous evolution of agriculture in these two centres, although the discovery of more ancient agricultural remains in this and other areas may require an alternate explanation. Recent archaeological evidence suggests that the evolution of agriculture was a gradual and complex process, probably concentrated in certain centres of innovation from which it spread to other areas, taking with it some cultigens, but also domesticating others along the way (Harris 1967).

Whatever its origins, the evolution of agriculture resulted in a profound change in the processes of human interference with the plant landscape. Henceforth, the process was no longer limited to mere interference with existing species ranges and vegetation composition, but expanded to the development of new domesticated varieties and species, the synthesis of entirely new plant assemblages composed of these cultigens, and extensive environmental modification creating a suite of new habitats.

Beginning in the mid-nineteenth century in western Europe and later spreading to other parts of the world, an accelerating technology added a new dimension to man's impact upon the earth's plant cover. This increased technological capacity has resulted in modification and extension of existing agricultural and pre-agricultural processes affecting the plant landscape, and the emergence of others as major contributors. Agricultural activities have been extended and modified by mechanization, pest suppression and environmental improvement, notably by fertilization. Evolution has continued to be channelled by the selective breeding of specialized cultigen varieties. Forestry, one of the last major vestiges of a hunting and gathering economy, has been dramatically extended and is only slowly being replaced by tree farming.

Industrial activity has created a range of new habitats for plants by modification of terrestrial and atmospheric environments. Extreme new terrestrial habitats have been created by building, highway construction, mining, etc. Similarly, the advent of atmospheric pollutants in high concentrations has created radically new environments for plant life. Today, the earth's plant cover continues to change rapidly as a result of these processes.

Environmental modification

The effect of man upon the earth's plant cover has been exercised for the most prolonged period of time through modification in various ways of the plant environment. The most pervasive form of this environmental modification has been the extension of open and disturbed habitats in the more densely vegetated areas of the earth. These open habitats, in contrast to the insulated forest micro-environment, experience high light levels, wide temperature fluctuations, are often covered by bare mineral soil and are temporarily free of intense competition. Under pristine conditions such open habitats existed only sporadically at tree fall sites, in river flood plains, landslip scars, wildfire sites, etc. The expansion of such habitats by man-induced fires, agriculture, and later, industrial activities has permitted both a massive range expansion of pre-adapted species, and the evolution of new genotypes through ecotypic diversification and by hybridization among previously isolated populations (Anderson 1956). These species have become the ruderal plants of the present landscape which, when bothersome to man, are regarded as weeds.

These ruderal species characteristically possess one or more specialized adaptations to the disturbed habitat. Most commonly these include short life cycles, ready vegetative spread, abundant small seed, and seed capable of prolonged viability in the soil. The history of some British weeds has been traced to the Late Glacial, and even Full Glacial, by Godwin (1956) using pollen records. At that time, these species existed in the open vegetation that preceded postglacial reforestation and later extended their ranges into the open habitats created by Neolithic man and his successors. J. D. Sauer (1952) has traced the expansion of pokeweed (*Phytolocca americana*) in eastern North America during the post-settlement period. This species, native to the bare mineral soil substrates of floodplains, proved suitably pre-adapted to agricultural land into which it spread. Other species have been elevated to weed status even more recently. The expansion of forestry in the Pacific Northwest has only recently elevated red alder (*Alnus rubra*, pl. 1) to weed status, as a consequence of its aggressive colonization of logged settings there.

Probably the most ancient means of extensive environmental modification by man has been through the use of fire, although considerable controversy surrounds the extent and importance of this. At one extreme, some have argued that most fires have been the result of human activity and that most grasslands are the product of these fires (e.g. C. O. H. Sauer 1950). However, although the occurrence of man-set fires has been widely documented (e.g. Stewart 1956; Bartlett 1967), there are also strong indications that wildfire incidence can be high (chapter 3). Also, palaeobotanical evidence suggests that grasslands were in existence prior to man's appearance; fossil evidence from the North American grassland area suggests that this formation was in existence there in the mid-Tertiary (Dix 1964). Similarly, Wymstra and Van der Hammen (1966) have found evidence, in a pollen sequence, of savanna grasslands in northern South America from the Late Glacial. However, in both situations, man-induced fires appear to have played some role in extending these grasslands. An extension of open grassland in northern South America during the last 3000 years is attributed by Wymstra and Van der Hammen to human activities. Similarly, the post-settlement extension of woody vegetation into prairie in areas such as Wisconsin (Curtis 1959) and the Canadian Prairies (Bird 1961) strongly suggests that Indian-set fires may have played some role in maintaining an extended periphery to this formation. Consequently, the central issue in the controversy over man-induced fires should concern the importance of this phenomenon relative to natural fires in *particular situations*: not the universal application of a single explanation.

In addition to his ancient use of fire, agricultural and industrial man has effected far-reaching changes in the environment of plants by other means. Agriculture brought with it soil erosion and the exposure of substrates for plant colonization that were rarely found before. Irrigated farming extended the ephemeral-pool habitat. Since the industrial revolution, the use of fertilizers in agriculture has extended areas of extreme soil fertility that were previously confined to middens and animal droppings. Herbicides have become a powerful new selective force in weedy plant evolution as has atmospheric pollution on all plants near industrial centres (Sinclair 1969). Mining and other industrial activities have created extensive new areas of raw mineral substrate, often of high toxicity. The study of plant colonization and evolution on these wastes is only in its infancy (Bradshaw 1970).

Together, these recent environmental modifications have resulted in a favouring of suitable pre-adapted genotypes capable of survival in the altered conditions, or able rapidly to evolve suitable adaptations. The result is an expanding ruderal or secondary vegetation cover.

Plant dissemination

Accompanying man's creation of new environments has been a remarkable dissemination of plants to new areas, across barriers that could not normally be crossed under pristine conditions. The dissemination has normally been intentional in the case of crops and accidental in the case of ruderals. The result has been an extensive homogenization of the flora of agricultural and disturbed habitats (e.g. Parsons 1972).

Accidental dissemination of ruderal species has been important in homogenizing the ruderal and weed vegetation of broadly similar climatic zones. At high elevations in the Blue Mountains of Jamaica, temperate weeds such as plantain (*Plantago*) and clover (*Triflolium*) can be found along roadsides, probably introduced in the eighteenth century as seed in hay used to feed horses at a military garrison in this area. Similarly, many weed floras contain a high proportion of widespread species, with few endemics (table 10.1). Although the homogeneity of most weed floras can be ascertained by a comparison of floras from different areas, it is frequently difficult to determine the areas of origin of the constituent species. A rapid spread before botanical collections were made and probable multiple introductions by several routes confuse the

Table 10.1 Geographic affinities (present range) of 183 weed species sampled in Belizean milpas

Range	No. of species	Percentage of total
Cosmopolitan	1	0.5
Pan-tropical (and subtropical)	56	30.6
Tropical America and Africa	11	6.0
Tropical America and Asia	2	1.1
Tropical America and Pacific Islands	6	3.3
Pan-America (tropical and subtropical)	57	31.1
Middle America (Caribbean and Gulf of Mexico rim)	21	11.5
Central America (mainland)	20	10.9
Belize and Yucatan	4	2.2
Unknown range	5	2.7

Source: Kellman and Adams (1970)

issue considerably. However, both the pollen record in areas occupied by man for prolonged periods (e.g. Godwin 1956) and rapid changes in recently settled areas (e.g. Clarke 1956) allow some assessment of the relative importance of indigenous and exotic weeds there. The persistence through time of many weed species in 'old' agricultural areas such as western Europe, and the preponderance of species from these areas in many of the more recently disturbed parts of the world suggests that prolonged exposure to agricultural man has favoured the evolution of successful weedy genotypes. In competition with these, only the most aggressive pioneers of local origin have been able to successfully establish in the weed flora of many areas. However, less severely disturbed waste places may contain a larger proportion of local ruderals.

While the widespread dissemination of aggressive ruderals has often presented a serious agricultural problem, dissemination of cultivated plants has frequently provided benefits far beyond the simple new resource that it introduced to other areas. Widespread plant movements have permitted the bringing together of previously isolated genotypes with consequent hybridization and diversification (see below). In addition the movement of crops to new areas in the post-Columbian period has permitted many to escape the predators and diseases present in their native habitats. For this reason, rubber production in the Amazon basin, native home of *Hevea brasiliensis,* has floundered because of infestation by South American leaf blight (*Dothidella ulei*). In contrast, the same species introduced blight-free into southeast Asia and quarantined since introduction has proved highly successful.

In sum, this wholesale dissemination of plants, when combined with the creation of uniformly disturbed habitats, has engendered an extensive homogenization of much of the world's vegetation. Moreover, the process continues to accelerate, producing a uniformity of world vegetation that has not existed since the early Tertiary. While this has promoted the evolution of many new species in the aggressive migrant flora, it has also led to a reduced range or extinction of many others unable to compete successfully in the new disturbed habitats and subsequently eliminated or confined to undisturbed environments (e.g. Egler 1942a).

Channelling of plant evolution

Organic evolution proceeds through two steps: the production of new genetic material (or genotypes) and the 'sifting' of this by natural selection. Man has impinged both intentionally and accidentally upon this process at both steps. New genotypes have been created by spontaneous hybridization and, more recently, by plant breeding and artificial mutagenesis, while new selective 'sieves' have been provided by environmental modification and selective propagation.

Under natural conditions, new genotypes are created by an accumulation of mutations and, occasionally, spontaneous hybridization and polyploidy. The preservation in a relatively undiluted form of those new genotypes that survive natural selection is favoured by ecological specialization in spatially segregated habitats. However, where such divergent, but still potentially interfertile genotypes are brought together, as in cultivated fields and waste places, the incidence of hybridization is markedly increased. This hybridization often involves a slow infiltration of genes between genotypes by back-crossing, a process termed 'introgressive hybridization' by Anderson (1949). The result of these processes is the appearance of new cultigen and ruderal genotypes with hybrid ancestry. Hybridization appears to have occurred between and within cultigen species, between and within weed species, and between weed and cultigen species. This has resulted in an exceedingly complex ancestry to many of our existing cultigens and weeds, and has made the identification of progenitors and pathways of evolution extremely difficult to determine. By detailed taxonomic and cytological examination and experimentation it has proven possible to identify the progenitors of some species such as the grain amaranths (J. D. Sauer 1950), but many others are of obscure ancestry.

The intentional production of new genotypes by cross-fertilization or artificially induced mutation is a more recent process confined to cultivated plants. It has been used most extensively in the production of new crop varieties suitable for specialized environments or varieties of extremely high productivity. In recent years, the most dramatic examples of the latter have been the so-called 'miracle' varieties of rice and wheat, created by intentionally crossing previously isolated varieties of these crops. The use of mutagens has been confined primarily to producing new ornamental genotypes.

The persistence of genetic changes depends upon the success of the plant possessing these in surviving and producing offspring. Under natural conditions, few changes survive the rigours of this natural selection process as the plant concerned usually succumbs to competition from more efficient genotypes. In contrast, new genotypes appearing in man-created habitats are often freed of competition and other rigours of natural selection. In waste places, colonizing ruderals may flourish because of the absence of competition. In cultivated fields, competition is removed by weeding, pests are often suppressed, and the abiotic environment improved by irrigation and fertilization. In such an environment, natural selection is at a minimum, and new genotypes can survive despite a low efficiency. Furthermore. those new genotypes thought by man to be desirable because of food value, utility or appearance, may be selectively propagated. In this way, natural selection pales as an evolutionary sifter and human selection predominates. The result has been the development in the pre-industrial period of many hundreds of cultivated plant species and varieties, some possessing bizarre, but culturally favoured, morphological features and incapable of survival without man's assistance. Probably the most notable example of this among common crops is corn (*Zea mays*) whose husk-enclosed seed depends upon man for release.

A notable feature of crop plant domestication is the paucity of recently domesticated crops. All major food crops and almost all other plant cultigens were domesticated in pre-history. Apart from ornamentals, rubber is one of the few examples of recent plant domestication. In contrast to this great diversification of cultigens in the pre-historic period, the post-industrial period has witnessed a reversion towards homogeneity of cultigen genotypes, reflecting the uniform crops required by highly technological modern agriculture and by world trade. The trend represents a serious threat to the resources of genetic variability built into cultigen stock over many millennia of cultivation and selection.

The earth's plant cover today

The earth's plant cover today is composed of three sorts of vegetation: agricultural, ruderal and remnant natural. Most flat or gently sloping areas with a suitable climate and fertile soil are covered by homogeneous artificial plant assemblages of low floristic diversity and often composed of uniform genetic stock. These are maintained relatively free of natural selective pressures by the use of mechanical cultivation, herbicides, insecticides, irrigation and fertilizers. While highly productive of a uniform product, this vegetation is extremely vulnerable and could not long survive without man's care. It lacks the variability, and consequent flexibility of response, of natural vegetation, and is composed of species of low competitive ability, which often require man's intervention for successful propagation.

Juxtaposed with these agricultural plant assemblages are extensive areas of ruderal or secondary vegetation, occupying the disturbed and waste places created by man's activities. The species composing this vegetation are usually aggressive ruderals drawn both from the local native flora and from a widely disseminated weedy flora. While the latter component provides a strong element of homogeneity in the flora at continental and world scales, the presence of a more endemic element lends localized character and often considerable floristic richness to this vegetation. As occupiers of recently created and often continually disturbed sites, these plant assemblages are almost always in a highly dynamic state.

Interspersed amongst this man-induced vegetation are increasingly smaller areas of natural or pristine vegetation, little affected by man and in a state which, while often far from steady, is far less dynamic than that of ruderal vegetation. While considerable attention in plant geography and ecology has been devoted to this pristine vegetation, little thought has been given to the possible consequences of its increasing insularity. If MacArthur and Wilson's (1967) model of insular biotas proves even partially correct, the fate of these islands of pristine vegetation could well involve appreciable impoverishment as immigration rates fall due to increasing isolation.

The vast bulk of plant geographic theory has grown out of studies on pristine or little disturbed natural vegetation. In view of the relative stability of this, it is not surprising that this theory has usually placed great emphasis upon the static relationship between plant species or assemblages and the environment. Dynamism, when considered, has usually been conceptualized as a directional progression to static end-points. While this body of theory has provided valuable insights, it is in itself inadequate to deal with the earth's present highly dynamic plant cover. Man-induced vegetation has received remarkably little attention. Although the specialized field of plant domestication has been the centre of research for many years, attention to ruderal vegetation has largely centred on weed control. Little attention has been devoted to the distribution and dynamics of weed assemblages or the other vast areas of ruderal vegetation covering the earth.

It would appear that this vegetation will require not only more research, but also a largely new body of theory and methods cognizant of its extreme dynamism. It is no accident that the long neglected field of plant population dynamics is now being actively developed by weed ecologists (Harper 1967), to whom this dynamism is most apparent. A major frontier of research in plant geography appears to lie in the study of the world's ruderal vegetation cover. Only by pursuing far more case studies of this can we hope to develop an adequate body of theory about it, an understanding of its form and function, and an ability to predict its variation in space and time.

Postscript

The future of the field

Whittaker (1957) has described recent changes in ecological concepts as '... the dissolution of a coherent, well-ordered, deductive system for the interpretation of vegetation, and its replacement by less interdependent inductive part-knowledges of different vegetational problems'. Although referring specifically to the dissolution of Clements's concepts, the generalization can be extended to much of the field of plant geography.

The earth's plant cover is now seen to be composed of species populations whose structures are seldom uniform and whose evolutionary history is rarely as narrowly channelled as once supposed. Environmental conditions impinging upon these species populations have been shown to be intricate and species population ranges rarely stable. Vegetation history increasingly reveals varied patterns of change and the presence of unfamiliar species assemblages in the past. Past human influences upon the earth's plant cover now appear to have been complex. In sum, the bases of deductive systems have been eroded by the complexity of detail in form and process now revealed in the earth's plant cover.

Recent research on the earth's plant cover reflects this methodological transformation. The frontiers of the field no longer comprise descriptions of vegetation in formalized modes. Instead, the frontiers lie in detailed treatments of the species populations that constitute this plant cover, their patterns, control mechanisms and modes of integration into intricate and continually evolving assemblages. Studies of plant population dynamics, species diversity and species packing have become centres of interest, as have experimental treatments of plant interaction mechanisms. Further developments in the field of plant population diffusion and more formal treatments of stochastic processes in vegetation may be anticipated.

In juxtaposition to this growing body of inductive analysis have been some attempts to redevelop synthetic systems of interpretation, exemplified by ecological energetics and models of diversity and insular colonization. Only time will reveal whether these attempts have been premature.

Implications for plant geographers

The demise of deductive systems for interpreting the earth's plant cover holds profound implications for the aspiring plant geographer. He can no longer function in the security of coherent systems with familiar ties to Davisian geomorphology, nor can he confine his endeavours to description alone. Instead enlightened interpretation of the patterns in the earth's plant cover requires an increased familiarity with the complex processes and mechanisms that operate within it. This requires not only an appreciation of the fundamentals of plant biology, but also a familiarity with the developments taking place within plant ecology, population biology and evolutionary theory. Failure to achieve such a familiarity will ultimately consign the plant geographer to the role of a maker of maps. Nor should the plant geographer see in the newly emerging deductive systems of interpretation a means of avoiding this requirement. The adoption of such systems without a sound appreciation of their biological basis can lead once more into the intellectual culs-de-sac from which the field is emerging at last.

The plant geographer willing to prepare himself in this way can find himself in an advantageous position. Established sciences are seldom free of institutionalization, and dogma abounds in the biological sciences no less than in other sciences. The enlightened outsider, uninhibited by the strictures of training within a conventional discipline, is free to ask questions and make comments that are inconceivable to the conventional practitioner. Moreover, there is much in the geographer's present training to prepare him for plant geography. Preparation in quantitative techniques, very useful in plant geography, is now happily an essential part of most geography curricula. A sound grounding in modern climatology, pedology and geomorphology is a valuable introduction to plant environments in the field. A familiarity with man and his activities is a useful preparation both for the analysis of disturbed vegetation and for the treatment of applied plant geographic problems.

However, conventional geography curricula emphasize an ability to think only in terms of economic, cultural and geophysical processes. If biogeography is to become a viable component of the field, it is essential to develop an ability to think also in terms of biological processes.

Bibliography

ABBOTT, H. G. and QUINK, T. F. (1970) Ecology of eastern white pine seed caches made by small forest mammals. *Ecol.* 51, 271—8.

AGNEW, A. D. Q. (1961) Ecology of *Juncus effusus* L. in North Wales. *J. Ecol.* 49, 83—102.

ALLEN, W. E. (1929) The problem of significant variables in natural environments. *Ecol.* 10, 223—7.

ANDERSON, A. J. B. (1971) Ordination methods in ecology. *J. Ecol.* 59, 713—26.

ANDERSON, E. (1949) *Introgressive hybridization.* New York.

ANDERSON, E. (1956) Man as a maker of new plants and new plant communities, in THOMAS, W. L. (ed.) *Man's role in changing the face of the earth.* Chicago, 763—77.

ANDERSON, H. G. (1969) Growth form and distribution of vine maple (*Acer circinatum*) on Mary's Peak, western Washington. *Ecol.* 50, 127—30.

ANDERSON, M. C. (1966) Ecological grouping of plants. *Nature* 212, 54—6.

ARMSTRONG, J. S. (1967) Derivation of theory by means of factor analysis or Tom Swift and his electric factor analysis machine. *Am. Stat.* 21 (5), 17—21.

ARNOLD, C. A. (1947) *An introduction to paleobotany.* New York.

AUBREVILLE, A. (1938) La forêt coloniale: les forêts de l'Afrique occidentale française. *Ann Acad. Scie colon.* 9, 1—245.

AXELROD, D. I. (1958) Evolution of the Madro-Tertiary geoflora. *Bot. Rev.* 24, 433—509.

BAKER, H. G. (1972) Seed weight in relation to environmental conditions in California. *Ecol.* 53, 997—1010.

BAKER, H. G. and STEBBINS, G. L. (eds.) (1965) *The genetics of colonizing species.* New York.

BARROW, M. D., COSTIN, A. B. and LAKE, P. (1968) Cyclical changes in an Australian fjaeldmark community. *J. Ecol.* 56, 89—96.

BARTLETT, H. H. (1967) *Fire in relation to primitive agriculture and grazing in the tropics; annotated bibliography.* Ann Arbor, Mich.

BATCHEIDER, R. B. and HIRT, H. F. (1966) *Fire in tropical forests and grasslands.* U.S. Army Natick Lab., Tech. Rep. 67—41—ES. Natick, Mass.

BAUR, G. N. (1968) *The ecological basis of rainforest management.* Sydney.

BEADLE, N. C. W. (1951) The misuse of climate as an indicator of vegetation and soils. *Ecol.* 32, 343—5.

BEARD, J. S. (1944) Climax vegetation in tropical America. *Ecol.* 25, 127—58.
BEARD, J. S. (1945) The Mora forests of Trinidad, British West Indies. *J. Ecol.* 33, 173—92.
BESCHEL, R. E. and WEBER, P. J. (1962) Gradient analysis in swamp forests. *Nature* 194, 207—9.
BILLINGS, W. D. (1938) The structure and development of old field shortleaf pine stands and certain associated physical properties of the soil. *Ecol. Monogr.* 8, 437—99.
BIRD, R. D. (1961) *Ecology of the aspen parkland of Western Canada in relation to land use.* Canada Dept. Agric., Res. Br. Publ. 1066. Ottawa.
BLACKMAN, G. E. and RUTTER, A. J. (1946) Physiological and ecological studies in the analysis of plant environment: I. The light factor in the distribution of bluebell (*Scilla non-scripta*) in woodland communities. *Ann Bot.* 10, 361—90.
BORMANN, F. H. (1953) The statistical efficiency of sample plot size and shape in forest ecology. *Ecol.* 34, 474—87.
BRADSHAW, A. D. (1970) Pollution and plant evolution. *New. Sci.* 48, 497—500.
BRAUN, E. L. (1950) *Decidous forests of eastern North America.* Philadelphia.
BRAUN-BLANQUET, J. (1932) *Plant sociology,* translated by G. D. Fuller and H. S. Conard. New York.
BRAY, J. R. and CURTIS, J. T. (1957) An ordination of the upland forest communities of southern Wisconsin. *Ecol. Monogr.* 27, 325—49.
BRAY, J. R. and GORHAM, E. (1964) Litter production in forests of the world. *Adv. in Ecol. Res.* 2, 101—57.
BROWN, R. T. and CURTIS, J. T. (1952) The upland conifer-hardwood forests of northern Wisconsin. *Ecol. Monogr.* 22, 217—34.
CHANEY, R. W. (1947) Tertiary centers and migration routes. *Ecol. Monogr.* 17, 139—48.
CHANEY, R. W. and AXELROD, D. I. (1959) *Miocene floras of the Columbia Plateau.* Carnegie Inst. Washington Publ. 617. Washington.
CHANEY, R. W. and SANBORN, E. I. (1933) *The Goshen flora of west central Oregon.* Carnegie Inst. Washington Publ. 439. Washington.
CLARK, A. H. (1956) The impact of exotic invasion on the remaining New World mid-latitude grasslands, in THOMAS, W. L. (ed.) *Man's role in changing the face of the earth.* Chicago, 737—62.
CLEMENTS, F. E. (1928) *Plant succession and indicators.* New York.
CLEMENTS, F. E. (1936) Nature and structure of the climax. *J. Ecol.* 24, 252—84.
COOMBE, D. E. (1957) The spectral composition of shade light in woodlands. *J. Ecol.* 45, 823—30.
COOPER, C. F. (1961) The ecology of fire. *Sci Am.* 204(4), 150—60.
COOPER, W. S. (1926) The fundamentals of vegetational change. *Ecol.* 7, 391—413.
COOPER, W. S. (1928) Seventeen years of successional change upon Isle Royale, Lake Superior. *Ecol.* 9, 1—5.
COTTAM, G. and CURTIS, J. T. (1949) A method for making rapid surveys of woodlands by means of pairs of randomly selected trees. *Ecol.* 30, 101—4.
COTTAM, G. and CURTIS, J. T. (1955) Correction for various exclusion angles in the random pairs method. *Ecol.* 36, 767.
COTTAM, G. and CURTIS, J. T. (1956) The use of distance measures in phytosociological sampling. *Ecol.* 37, 451—60.

COWLES, H. C. (1899) Ecological relations of the vegetation on the sand dunes of Lake Michigan. *Bot. Gaz.* 27, 95—117, 167—202, 281—308, 361—91.

CRAIG, A. J. (1972) Pollen influx to laminated sediments: a pollen diagram from northeastern Minnesota. *Ecol.* 53, 46—57.

CRAWFORD, R. M. M. and WISHART, D. (1968) A rapid classification and ordination method and its application to vegetation mapping, *J. Ecol.* 56, 385—404.

CROCKER, R. L. and MAJOR, J. (1955) Soil development in relation to vegetation and surface age in Glacier Bay, Alaska. *J. Ecol.* 43, 427—48.

CURTIS, J. T. (1959) *The vegetation of Wisconsin.* Madison.

CURTIS, J. T. and McINTOSH, R. P. (1951) An upland forest continuum in the prairie-forest border region of Wisconsin. *Ecol.* 32, 476—96.

DAGNELIE, P. (1960) Contribution à l'étude des communautés végétales par l'analyse factorielle. *Bull. Serv. Carte phytogeogr.* Ser. B. 5, 7—71, 93—195.

DANSEREAU, P. (1958) *A universal system for recording vegetation.* Contr. de l'Inst. Bot. de l'Univ. de Montreal 72. Montreal.

DANSEREAU, P. and LEMS, K. (1957) *The grading of dispersal types in plant communities and their ecological significance.* Contre. de l'Inst. Bot. de l'Univ. de Montreal 71. Montreal.

DARLINGTON, H. T. (1931) The 50-year period for Dr Beal's seed viability experiment. *Am. J. Bot.* 18, 262—5.

DARWIN, C. (1859) *On the origin of species by means of natural selection, or the preservation of favoured species in the struggle for life.* London.

DAUBENMIRE, R. F. (1959) *Plants and environment.* 2nd edn. New York.

DAVIS, M. B. (1967) Late-glacial climate in northern United States: a comparison of New England and the Great Lakes region, in CUSHING, E. J. and WRIGHT, H. E. (eds.) *Quaternary paleoecology.* New Haven, 11—43.

DAVIS, T. A. W. and RICHARDS, P. W. (1933—4) The vegetation of Moraballi Creek, British Guiana: an ecological study of a limited area of tropical rain forest. Parts I and II. *J. Ecol.* 21, 350—84; 22, 106—55.

DAVIS, W. M. (1899) The geographical cycle. *Geogr. J.* 14, 481—504.

DAWKINS, H. C. (1959) The volume increment of natural tropical high-forest and limitations on its improvements. *Emp. For. Rev.* 38, 175—80.

DEL MORAL, R. and CATES, R. G. (1971) Allelopathic potential of the dominant vegetation of western Washington. *Ecol.* 52, 1030—7.

DE VRIES, D. M. (1953) Objective combinations of species. *Acta bot. neerl.* 1, 497—9.

DIMBLEBY, G. W. (1961) Soil pollen analysis. *J. Soil Sci.* 12, 1—11.

DIX, R. L. (1964) A history of biotic and climatic changes within the North American grassland, in CRISP, D. J. (ed.) *Grazing in terrestrial and marine environments.* Brit. Ecol. Soc. Symp. 4. Oxford, 71—89.

DUNBAR, M. J. (1968) *Ecological development in polar regions.* Englewood Cliffs, N.J.

EDWARDS, A. W. F. and CAVALLI-SFORZA, L. L. (1965) A method for cluster analysis. *Biometrics* 21, 362—75.

EGLER, F. E. (1942a) Indigene versus alien in the development of arid Hawaiian vegetation. *Ecol.* 23, 14—23.

EGLER, F. E. (1942b) Vegetation as an object of study. *Phil. of Sci.* 9, 245—60.

EGLER, F. E. (1954) Vegetation science concepts: I. Initial floristic composition, a factor in old field vegetation development. *Veg.* 4, 412—17.

EIS, S. (1962) *Statistical analysis of several methods for estimation of forest habitats and tree growth near Vancouver, British Columbia.* Univ. of Brit. Columbia, Fac. of For., Bull. 14. Vancouver.

ERRINGTON, J. C. (1973) The effect of regular and random distributions on the analysis of pattern. *J. Ecol.* 61, 99—105.

EVANS, F. C. (1956) Ecosystem as the basic unit in ecology. *Sci.* 123, 1127—8.

EVERITT, B. L. (1968) Use of cottonwood in an investigation of recent history of a flood plain. *Am. J. Sci.* 266, 417—39.

FAEGRI, K. and IVERSEN, J. (1964) *Textbook of pollen analysis.* 2nd edn. New York.

FLENLEY, J. R. (1967) *The present and former vegetation of the Wabag region of New Guinea.* Unpubl. Ph.D. thesis, Aust. Nat. Univ. Canberra.

FLENLEY, J. R. (1969) The vegetation of the Wabag Region, New Guinea highlands: a numerical study. *J. Ecol.* 57, 465—90.

FRITTS, H. C., SMITH, D. G., CARDIS, J. W. and BUDELSKY, C. A. (1965) Tree-ring characteristics along a vegetation gradient in northern Arizona. *Ecol.* 46, 393—401

GARMAN, E. H. (1951) *Seed production by conifers in the coastal region of British Columbia related to dissemination and regeneration.* Brit. Columbia For. Serv., Res. Div., Tech Publ. T 35. Victoria.

GATES, D. M. (1968) Toward understanding ecosystems. *Adv. in Ecol. Res.* 5, 1—35.

GIDDINGS, J. L. (1941) *Dendrochronology of northern Alaska.* Univ. of Alaska Publ. 5. College.

GLEASON, H. A. (1926) The individualistic concept of the plant association. *Bull. Torrey Bot. Club* 53, 7—26.

GLEASON, H. A. (1939) The individualistic concept of the plant association. *Am. Midl. Nat.* 21, 92—110.

GODWIN, H. (1956) *The history of the British flora.* Cambridge.

GOOD, N. F. and GOOD, R. E. (1972) Population dymamics of tree seedlings and saplings in a mature eastern hardwood forest. *Bull Torrey Bot. Club* 99, 172—8.

GOODALL, D. W. (1953) Objective methods for the classification of vegetation: I. The use of positive inter-specific correlation. *Aust. J. Bot.* 1, 39—63.

GOODALL, D. W. (1954) Objective methods for the classification of vegetation: III. An essay in the use of factor analysis. *Aust, J. Bot.* 2, 304—24.

GOODALL, D. W. (1963) The continuum and the individualistic association. *Veg.* 11, 297—316.

GOWER, J. C. (1966) Some distance properties of latent roots and vector methods in multivariate analysis. *Biometrika* 53, 325—38.

GREIG-SMITH, P. (1952) The use of random and contiguous quadrats in the study of the structure of plant communities. *Ann. Bot.* 16, 293—318.

GREIG-SMITH, P. (1961) Data on pattern within plant communities: I. The analysis of pattern. *J. Ecol.* 49, 695—702.

GREIG-SMITH, P. (1964) *Quantitative plant ecology.* 2nd edn. London.

GROSENBAUGH, L. R. (1952) Plotless timber estimates — new, fast, easy. *J. For.* 50, 32—7.

GUPPY, H. B. (1917) *Plants, seeds and currents in the West Indies and Azores.* London.

HADDOCK, P. G., WALTERS, J. and KOZAK, A. (1967) *Growth of coastal and interior provenances of Douglas-fir (Pseudotsuga menziesii (Mirb.) Franco)* at *Vancouver and Haney in British Columbia.* Univ. of Brit. Columbia, Fac. of For., Res. Pap. 79. Vancouver.

HARLAN, J. R. (1961) Geographic origin of plants useful to agriculture, in HODGSON, R. E. (ed.) *Germ plasm resources.* Am. Assoc. Adv. Sci. publ. 66. Washington, 3—19.

HARPER, J. L. (1967) A Darwinian approach to plant ecology. *J. Ecol.* 55, 247—70.

HARPER, J. L., LOVELL, P. H. and MOORE, K. G. (1970) The shapes and sizes of seeds. *Ann. Rev. Ecol. System.* 1, 327—56.

HARRIS, D. R. (1965) *Plants animals and man in the Outer Leeward Islands, West Indies.* Berkeley.

HARRIS, D. R. (1967) New light on plant domestication and the origins of agriculture: a review. *Geogr. Rev.* 57, 90—107.

HARTLEY, C. W. S. (1968) The soil relations and fertilizer requirements of some permanent crops in West and Central Africa, in MOSS, R. P. (ed.) *The soil resources of tropical Africa.* Cambridge, 155—83.

HASTINGS, J. R. and TURNER, R. M. (1965) *The changing mile: an ecological study of vegetation change with time in the lower mile of an arid and semiarid region.* Tucson.

HELBAEK, H. (1964) First impressions of the Çatal Hüyük plant husbandry. *Anatolian Stud.* 14, 121—3.

HETT, J. M. (1971) A dynamic analysis of age in sugar maple seedlings. *Ecol.* 52, 1071—4.

HOPKINS, B. (1955) The species-area relations of plant communities. *J. Ecol.* 43, 409—26.

HUMBOLDT, A. VON (1807) *Ideen zu einer Geographie der Pflanzen nebst einen Naturgamalde der Tropenlander.* Tubingen.

HURLEBERT, S. H. (1969) A coefficient of inter-specific association. *Ecol.* 50, 1—9

JANZEN, D. H. (1970) Herbivores and the number of tree species in tropical forests. *Am. Nat.* 104, 501—28.

JANZEN, D. H. (1971) Seed predation by animals. *Ann. Rev. Ecol. System.* 2, 465—92.

JOHNSON, F. S. (1954) The solar constant. *J. Met.* 11, 431—9.

KELLMAN, M. C. (1969a) A critique of some geographical approaches to the study of vegetation. *Prof. Geogr.* 21, 11—14.

KELLMAN, M. C. (1969b) Plant species interrelationships in a secondary succession in coastal British Columbia. *Syesis* 2, 201—12.

KELLMAN, M. C. (1970a) On the nature of secondary plant succession. *Proc. Can. Assoc. Geogr., June 1970.* Winnipeg, 193—8.

KELLMAN, M. C. (1970b) *Secondary plant succession in tropical Montane Mindanao.* Canberra.

KELLMAN, M. C. (1970c) The influence of accessibility on the composition of vegetation. *Prof. Geogr.* 22, 1—4.

KELLMAN, M. C. (1973) Dry season weed communities in the upper Belize Valley. *J. Appl. Ecol.* 10, 683—94.

KELLMAN, M. C. (1974) The viable weed seed content of some tropical agricultural soils. *J. Appl. Ecol.* 11 (in press).

KELLMAN, M. C. and ADAMS, C. D. (1970) Milpa weeds of the Cayo District, Belize (British Honduras). *Can. Geogr.* 14, 323—43.

KENOYER, L. A. (1934) Forest distribution in southwestern Michigan as interpreted from the original land survey (1826–32). *Pap. Mich. Acad. Sci. Arts & Letters* 19, 107–11.

KERSHAW, K. A. (1960) *The detection of pattern and association. J. Ecol.* 48, 233–43.

KERSHAW, K. A. (1961) Association and co-variance analysis of plant communities. *J. Ecol.* 49, 643–54.

KERSHAW, K. A. (1963) Pattern in vegetation and its causality. *Ecol.* 44, 377–88.

KOMAREK, E. V. (1964) The natural history of lightning. *Proc. 3rd Tall Timbers Fire Ecol. Conf.* Talahassee, 139–84.

KUCHLER, A. W. (1949) A physiognomic classification of vegetation. *Ann. Assoc. Am. Geogr.* 39, 201–10.

LAMBERT, J. M. and DALE, M. B. (1964) The use of statistics in phytosociology. *Adv. in Ecol. Res.* 2, 59–99.

LINDEMAN, R. L. (1942) The trophic-dynamic aspect of ecology. *Ecol.* 23, 399–418.

LIVINGSTON, R. B. and ALLESSIO, M. L. (1968) Buried viable seed in successional field and forest stands, Harvard Forest, Massachusetts. *Bull. Torrey Bot. Club* 95, 58–69.

LIVINGSTONE, D. A. (1968) Some interstadial and postglacial pollen diagrams from eastern Canada. *Ecol. Monogr.* 38, 87–125.

LOUCKS, O. L. (1962) Ordinating forest communities by means of environmental scalars and phytosociological indices. *Ecol. Monogr.* 32, 137–66.

LÖVE, A. (1962) The biosystematic species concept. *Preslia* 34, 127–39.

MacARTHUR, R. H. and WILSON, E. O. (1967) *The theory of island biogeography.* Princeton.

McNAUGHTON, S. J. (1968) Autotoxic feedback in relation to germination and seedling growth in *Typha latifolia. Ecol.* 49, 367–9.

MacNAUGHTON-SMITH, P., WILLIAMS, W. T., DALE, M. B. and MOCKETT, L. G. (1964) Dissimilarity analysis: a new technique of hierarchical sub-division. *Nature* 202, 1034–5.

MacNEISH, R. S. (1965) The origins of American agriculture. *Antiq.* 39, 87–94.

MAJOR, J. (1958) Plant ecology as a branch of botany. *Ecol.* 39, 352–62.

MAJOR, J. (1961) Use in plant ecology of causation, physiology and a definition of vegetation. *Ecol.* 42, 167–9.

MAJOR, J. and PYOTT, W. T. (1966) Buried, viable seeds in two California bunchgrass sites and their bearing on the definition of a flora. *Veg.* 13, 253–82.

MARBUT, C. F. (1927) A scheme for soil classification. *Proc. 1st Internat. Congr. Soil Sci.* 4, 1–31.

MARTIN, P. S. (1963) Paleoclimatology and a tropical pollen profile. *Rep. 6th Congr. Internat. Assoc. Quat. Res., 1961* 2, 319–23. Warsaw.

MARTIN, P. S. and WRIGHT, H. E. (eds.) (1967) *Pleistocene extinctions: the search for a cause.* New Haven.

MASON, H. L. and LANGENHEIM, J. H. (1957) Language analysis and the concept *environment. Ecol.* 38, 325–40.

MATALAS, N. C. (1962) Statistical properties of tree ring data. *Bull. Internat. Assoc. Sci. Hydrol.* 7 (2), 39–47.

MAYER, A. M. and POLJAKOFF-MAYBER, A. (1963) *The germination of seeds.* London.

MIKESELL, M. W. (1969) The deforestation of Mount Lebanon. *Geogr. Rev.* 59, 1–28.

MINDEMAN, G. (1968) Addition, decomposition and accumulation of organic matter in forests. *J. Ecol.* 56, 355–62.

MOLINIER, R. and MULLER, P. (1938) La dissémination des espèces végètales. *Rev. Gen. de Bot.* 50, 53–72, 152–69, 202–21, 277–93, 341–58, 397–414, 427–88, 533–46, 598–614, 649–70.

MULLER, C. H. (1969) Allelopathy as a factor in ecological processes. *Veg.* 18, 348–57.

NEWBOULD, P. J. (1967) *Methods of estimating the primary production of forests.* Internat. Biol. Progr. Hdbk. 2. Oxford.

NYE, P. H. and GREENLAND, D. J. (1960) *The soil under shifting cultivation.* Commonw. Bur. Soils Tech. Communic. 51. Farnham, Bucks.

OGAWA, H., YODA, K., OGINO, K. and KIRA, T. (1965) Comparative ecological studies on three main types of forest vegetation in Thailand: II. Plant biomass. *Nature and Life in S.E. Asia* 4, 49–81.

OLMSTED, C. E. (1944) Growth and development of range grasses: IV. Photoperiodic responses in twelve geographic strains of side-oats grama. *Bot. Gaz.* 106, 46–74.

OLMSTED, N. W. and CURTIS, J. D. (1947) Seeds of the forest floor. *Ecol.* 28, 49–52.

OLSON, J. S. (1958) Rates of succession and soil changes on southern Lake Michigan sand dunes. *Bot. Gaz.* 119, 125–70.

OLSON, S. L. and BLUM, K. E. (1968) Avian dispersal of plants in Panama. *Ecol.* 49, 565–6.

ORLOCI, L. (1966) Geometric models in ecology: I. The theory and applications of some ordination methods. *J. Ecol.* 54, 193–215.

ORLOCI, L. (1967) An agglomerative method for classification of plant communities. *J. Ecol.* 55, 193–206.

PARSONS, J. J. (1960) 'Fog drip' from coastal stratus, with special reference to California. *Weather* 15 (2), 58–62.

PARSONS, J. J. (1972) Spread of African grasses to the American tropics. *J. Ra. Mgmt.* 25, 12–17.

PELTON, J. (1953) Ecological life cycle of seed plants. *Ecol.* 34, 619–28.

POORE, M. E. D. (1955–6) The use of phytosociological methods in ecological investigations: I. The Braun-Blanquet system; II. Practical issues involved in an attempt to apply the Braun-Blanquet system; III. Practical applications; IV. General discussion of phytosociological problems. *J. Ecol.* 43, 226–44, 245–69, 606–51; 44, 28–50.

POORE, M. E. D. (1962) The method of successive approximation in descriptive ecology. *Adv. in Ecol. Res.* 1, 35–68.

POORE, M. E. D. (1963) Integration in the plant community, in *Jubilee Symp. Brit. Ecol. Soc.* Oxford, 213–26.

PUTWAIN, P. D. and HARPER, J. L. (1970) Studies in the dynamics of plant populations: III. The influence of associated species on populations of *Rumex acetosa* L. and *R. acetosella* L. in grassland. *J. Ecol.* 58, 251–64.

PUTWAIN, P. D., MACHIN, D. and HARPER, J. L. (1968) Studies in the dynamics of plant populations: II. Components and regulation of a natural population of *Rumex acetosella* L. *J. Ecol.* 56, 421–31.

RANKIN, H. T. and DAVIS, D. E. (1971) Woody vegetation in the Black Belt prairie of Montgomery County, Alabama, in 1845–46. *Ecol.* 52, 716–19.

RAUNKIAER, C. (1934) *The life form of plants,* translated by H. Gilbert-Carter. Oxford.

RAUP, H. M. (1942) Trends in the development of geographic botany. *Ann. Assoc. Am. Geogr.* 32, 319–54.

RIDLEY, H. N. (1930) *The dispersal of plants throughout the world.* Ashford, Kent.

ROBERTS, H. A. (1970) Viable weed seeds in cultivated soil. *Rep. Nation. Veg. Res. Stn. for 1969*, 25–38.

ROBERTSON, G. W. (1966) The light composition of solar and sky spectra available to plants. *Ecol.* 47, 640–3.

RODIN, L. E. and BRAZILEVICH, N. I. (1967) *Production and mineral cycling in terrestrial vegetation*, translated by G. E. Fogg. Edinburgh.

RUSSELL, E. J. (1961) *Soil conditions and plant growth* 9th edn. London.

ST JOSEPH, J. K. S. (ed.) (1966) *The uses of air photography; nature and man in a new perspective.* New York.

SALISBURY, E. J. (1942) *The reproductive capacity of plants.* London.

SAUER, C. O. (1950) Grassland, climax fire, and man. *J. Ra. Mgmt.* 3, 16–21.

SAUER, C. O. (1952) *Agricultural origins and dispersals.* New York.

SAUER, J. D. (1950) The grain Amaranths: survey of their history and classification. *Ann. Miss. Bot. Gdn.* 37, 561–632.

SAUER, J. D. (1952) A geography of pokeweed. *Ann. Miss. Bot. Gdn.* 39, 113–25.

SAUER, J. D. (1967) *Plants and man on the Seychelles coast; a study in historical biogeography.* Madison.

SAUER, J. D. (1969) Oceanic islands and biogeographical theory: a review. *Geogr. Rev.* 59, 582–93.

SCHIMPER, A. F. W. (1903) *Plant geography upon a physiological basis*, translated by W. R. Fisher. Oxford.

SCOTT, G. A. M. (1965) The shingle succession at Dungeness. *J. Ecol.* 53, 21–31.

SHINN, T. L. (1971) *Patterns of regeneration in old growth cedar-hemlock forest in coastal British Columbia.* Unpubl. M.A. thesis, Univ. of California. Berkeley.

SIMBERLOFF, D. S. and WILSON, E. O. (1970) Experimental zoogeography of islands. A two-year record of colonization. *Ecol.* 51, 934–7.

SINCLAIR, W. A. (1969) Polluted air: potent new selective force in forests. *J. For.* 67, 305–9.

SKELLAM, J. G. (1952) Studies in statistical ecology: I. Spatial pattern. *Biometrika* 39, 346–62.

SPOMER, G. G. (1973) The concept of 'interaction' and 'operational environment' in environmental analyses. *Ecol.* 54, 200–4.

STEBBINS, G. L. (1950) *Variation and evolution in plants.* New York.

STEBBINS, G. L. (1971) *Processes of organic evolution.* 2nd edn. Englewood Cliffs, N.J.

STEPHENS, S. G. (1958) Salt water tolerance of seeds of *Gossypium* species as a possible factor in seed dispersal. *Am. Nat.* 92, 83–92.

STEWART, O. C. (1956) Fire as the first great force employed by man, in THOMAS, W. L. (ed.). *Man's role in changing the face of the earth.* Chicago, 115–33.

STODDART, D. R. (1965) Geography and the ecological approach. The ecosystem as a geographical principle and method. *Geogr.* 50, 242–51.

STOKES, M. A. and SMILEY, T. L. (1968) *An introduction to tree-ring dating.* Chicago.

SWAN, J. M. A., DIX, R. I. and WHERHAHN, C. F. (1969) An ordination technique based on the best possible stand-defined axes and its application to vegetational analysis. *Ecol.* 50, 206—12.

SYMINGTON, C. F. (1933) The study of secondary growth on rainforest sites in Malaya. *Malay. For.* 2, 107—17.

TANSLEY, A. G. (1920) The classification of vegetation and the concept of development. *J. Ecol.* 8, 118—49.

TANSLEY, A. G. (1935) The use and abuse of vegetational concepts and terms. *Ecol.* 16, 284—307.

TARRANT, R. F., LU, K. C., BOLLEN, W. B. and FRANKLIN, J. F. (1969) *Nitrogen enrichment of two forest ecosystems by red alder.* U.S.D.A. For. Serv. Res. Pap. PNW—76. Portland, Ore.

TAUBER, H. (1967) Differential pollen dispersion and filtration, in CUSHING, E. J. and WRIGHT, H. E. (eds.) *Quaternary paleoecology.* New Haven, 131—41.

THOMPSON, H. R. (1958) The statistical study of plant distribution patterns using a grid of quadrats. *Aust. J. Bot.* 6, 322—43.

TURESSON, G. (1922) The genotypical response of the plant species to the habitat. *Hereditas* 3, 211—350.

VAN DER PIJL, L. (1972) *Principles of dispersal in higher plants.* 2nd edn. New York.

VAN STEENIS, C. G. G. J. (1967) Notes on the introduction of *Crassocephalum crepidioides* (Bth.) S. Moore in Indo-Australia. *J. Ind. Bot. Soc.* 46, 463—9.

VAVILOV, N. I. (1951) *The origin, variation, immunity and breeding of cultivated plants,* selected writings translated by K. S. Chester. Waltham, Mass.

VÉZINA, P. E. and BOULTER, D. W. K. (1966) The spectral composition of near ultra-violet and visible radiation beneath forest canopies. *Can. J. Bot.* 44, 1267—84.

VUILLEUMIER, B. S. (1971) Pleistocene changes in the fauna and flora of South America. *Sci.* 173, 771—80.

WACE, N. M. (1967) The units and uses of biogeography. *Aust. Geogr. Stud.* 5, 15—29.

WALKER, D. (1970a) Direction and rate in some British post-glacial hydroseres, in WALKER, D. and WEST, R. G. (eds.). *Studies in the vegetation history of the British Isles.* Cambridge, 117—39.

WALKER, D. (1970b) The changing vegetation of the montane tropics. *Search* 1, 217—21.

WARMING, E. (1909) *Oecology of plants,* translated by P. Groom and I. B. Balfour. Oxford.

WASSINK, E. C. (1953) Specification of radiant flux and radiant flux density in the irradiation of plants with artificial light. *J. Hort. Sci.* 28, 177—84.

WATT, A. S. (1947) Pattern and process in the plant community. *J. Ecol.* 35, 1—22.

WATTS, W. A. (1970) The full-glacial vegetation of northwestern Georgia. *Ecol.* 51, 17—33.

WATTS, W. A. (1971) Postglacial and interglacial vegetation history of southern Georgia and central Florida. *Ecol.* 52, 676—90.

WEAVER, J. E. and CLEMENTS, F. R. (1938) *Plant ecology.* 2nd edn. New York.

WEST, N. E. and CHILCOTE, W. W. (1968) *Senecio sylvaticus* in relation to Douglas-fir clear-cut succession in the Oregon Coast Range. *Ecol.* 49, 1101–7.

WEST, R. G. (1963) Inter-relations of ecology and Quaternary paleobotany, in *Jubilee Symp. Brit. Ecol. Soc.* Oxford, 47–57.

WESTLAKE, D. F. (1963) Comparisons of plant productivity. *Biol. Rev.* 38, 385–425.

WHITTAKER, R. H. (1953) A consideration of the climax theory: the climax as a population and pattern. *Ecol. Monogr.* 23, 41–78.

WHITTAKER, R. H. (1956) Vegetation of the Great Smoky Mountains. *Ecol. Monogr.* 26, 1–80.

WHITTAKER, R. H. (1957) Recent evolution of ecological concepts in relation to the eastern forests of North America. *Am. J. Bot.* 44, 197–206.

WHITTAKER, R. H. (1967) Gradient analysis of vegetation. *Biol. Rev.* 42, 207–64.

WHITTAKER, R. H. and FEENEY, P. P. (1971) Allelochemics: chemical interactions between species. *Sci.* 171, 757–70.

WHITTAKER, R. H. and NIERING, W. A. (1965) Vegetation of the Santa Catalina Mountains, Arizona: a gradient analysis of the south slope. *Ecol.* 46, 429–52.

WILBANKS, T. J. and SYMANSKI, R. (1968) What is systems analysis? *Prof. Geogr.* 20, 81–5.

WILDE, S. A. (1954) Floristic analysis of ground cover vegetation by a rapid chain method. *J. For.* 52, 499–502.

WILLIAMS, W. T. and LAMBERT, J. M. (1959) Multivariate methods in plant ecology: I. Association-analysis in plant communities. *J. Ecol.* 47, 83–101.

WILLIAMS, W. T. and LAMBERT, J. M. (1960) Multivariate methods in plant ecology: II. The use of an electronic digital computer for association-analysis. *J. Ecol.* 48, 689–710.

WILLIAMS, W. T. and LAMBERT, J. M. (1961a) Multivariate methods in plant ecology: III. Inverse association-analysis. *J. Ecol.* 49, 717–29.

WILLIAMS, W. T. and LAMBERT, J. M. (1961b) Nodal analysis of associated populations. *Nature* 191, 202.

WILLIAMS, W. T., LAMBERT, J. M. and LANCE, G. N. (1966) Multivariate methods in plant ecology: V. Similarity analyses and information-analysis. *J. Ecol.* 54, 427–45.

WOODWELL, G. M., WURSTER, C. F. and ISAACSON, P. A. (1967) DDT residues in an east coast estuary: a case of biological concentration of a persistent insecticide. *Sci.* 156, 821–4.

WRIGHT, H. E. (1968) The roles of pine and spruce in the forest history of Minnesota and adjacent areas. *Ecol.* 49, 937–55.

WYMSTRA, T. A. and VAN DER HAMMEN, T. (1966) Palynological data on the history of tropical savannas in northern South America. *Overdruk uit Leidse Geologische Mededelingen* 38, 71–90.

YARRANTON, G. A. (1967) Organismal and individualistic concepts and the choice of methods of vegetation analysis. *Veg.* 15, 113–6.

YARRANTON, G. A. (1969) Pattern analysis by regression. *Ecol.* 50, 390–5.

ZINKE, P. J. (1962) The pattern of influence of individual forest trees on soil properties. *Ecol.* 43, 130–3.

Glossary of terms used

Adsorbed ion   In soil, an ion held to the surface of a soil particle by electrostatic forces, but able to be removed in an exchange reaction.

Adventitious root   A root arising in an abnormal position (e.g. from the stem).

Aerobic   Active only in the presence of free oxygen.

Algae   Simple, photosynthetic plants with unicellular organs of reproduction.

Anion   A negatively charged ion.

Autotrophic   Capable of synthesizing organic substances from inorganic material; in higher plants by photosynthesis.

Carbohydrate   Organic compounds composed of carbon, hydrogen and oxygen. In plants especially cellulose and starch, which form the main structural components and food storage, respectively.

Cation   A positively charged ion.

Chlorophyll   Green pigment present in algae and autotrophic higher plants, and essential to the photosynthetic reaction.

Chromosome   Thread-shaped bodies composed largely of DNA (deoxyribonucleic acid) and protein, present in the nucleus of cells, and carrying the genetic code between generations. In higher plants, chromosomes normally exist in pairs, one derived from each parent plant.

Comose seed   Seed bearing one or more tufts of hair that aid in seed dispersal.

Crash   Precipitous decline in numbers of a population (normally animal).

Culm   The stem of a grass plant.

Cultigen   A cultivated plant genotype (also *cultivar*).

Diatom   A class (Bacillariophyceae) of unicellular of colonial algae with silicified skeletons.

Endosperm   Nutritive tissue surrounding and nourishing the plant embryo in seeds.

Enzyme   Complex organic substances acting as catalysts in chemical reactions within organisms.

Epidermis   Outermost layer of cells of an organism. In plants, a unicellular layer normally covered by a protective waxy cuticle that reduces moisture loss.

Flora   The total assemblage of plant species present in some area.

Forb   A herb other than grass.

Fungus   Member of the Mycophyta. Lower plants, lacking photosynthetic tissue, and subsisting upon dead and decaying organic matter.

Gene   A unit of the material of inheritance. Chemically, DNA (deoxyribonucleic acid) with varying arrangements of nucleotides, situated in chromosomes.

Gene ,pool   The pool of genetic variability present in an inter-breeding population of organisms.

Genotype   The genetic constitution of an organism (cf. *phenotype*).

Habitat   The specific kind of environment occupied by the individuals of a species.

Hybridization   Crossing between a genetically unlike pair of organisms, usually different species. The resulting hybrid may be fertile or sterile.

Inflorescence   A flowering shoot.

Insolation   Short-wave solar radiation.

Ion   An electrically charged atom or group of atoms.

'Knees'   Thick vertical projections from the roots of *Taxodium* grown in swampy conditions, apparently promoting aeration.

Layering   Vegetative propagation of plants from unspecialized trailing stems that are in contact with the ground.

Metabolism   Chemical processes occurring within an organism.

Niche   In general, the functional, spatial or temporal position of an organism within an interacting biotic assemblage.

Orthogonal   Right-angled.

Osmosis   Spontaneous differential diffusion of a solvent across a semi-permeable membrane, from solutions of high concentration to those of low concentration.

Perenniating bud   The overwintering undeveloped shoot of a perennial plant, consisting of a short stem and tightly packed immature leaves.

Phenotype   The sum of characteristics manifested by an organism. Normally some environmentally-induced phenotypic plasticity of organisms is possible within the limits set by their genotypes.

Phloem   Vascular tissue that conducts synthesized foods, and some mineral ions, through the plant (cf. xylem).

Photoperiodism   Response of plants to daylength (e.g. flowering in many species).

Photosynthesis   Synthesis of organic compounds from water and carbon dioxide using energy absorbed by chlorophyll from sunlight.

Phototropism   Light-stimulated directional plant growth.

Pneumatophore   Specialized vertical root branches, produced by some plants grown in swampy conditions, and containing respiratory tissue.

Pollen   Microspore of seed plants containing a greatly reduced male gamete. Movement of pollen between plants enables cross-fertilization.

Polyploidy   Production of new plant genotypes by the doubling of chromosome numbers.

Relict   A plant growing in an environment where it can no longer reproduce itself by seed.

Respiration   In plants, the breakdown of organic compounds using oxygen, to derive energy for metabolic processes.

Rhizome   Specialized underground stem, permitting perenniation and vegetative propagation.

Ruderal   A plant characteristic of waste places.

Soil colloid   Soil particles (mineral or organic) in a finely divided state, whose large surface areas provide most of the ion adsorption sites in soils.

Spore   Microscopic detached reproductive body of lower plants and some protozoa.

Stolon   Specialized horizontally growing plant stem, usually promoting extensive vegetative propagation.

Stoma   Pore on the epidermis of plants (especially leaves), through which gaseous exchange takes place and water vapour is lost in transpiration.

Sucker   A shoot arising below ground.

Synergism   Cooperative action of two or more factors to produce a net effect that is greater than the sum of each factor's individual effect when operating independently.

Taxon   A taxonomic group of any rank.

Translocation   Transport of materials within a plant, normally via xylem and phloem conducting tissue.

Trichome   Single or multi-celled outgrowth from an epidermal cell.

Vegetative propagation   Reproduction of new plants asexually, by suckering from roots, layering, etc.

Vivipary   Germination of a seed while still in a fruit and often while this remains attached to the parent plant.

Weed   Any plant considered undesirable by man. The adjective 'weedy' is often more loosely used to refer to any plant of ruderal habit.

Xylem   Vascular tissue by which water and mineral ions are conducted from roots to other parts of a plant (cf. phloem).

Index

1.   Pure stands of red alder (*Alnus rubra*) of two ages and densities, coastal British Columbia. Saplings in the foreground are 3 years old; mature trees in the background are approximately 30 years old. This species does not regenerate in its own shade and forms 'one generation' successional stands. (Photograph by Ian Hutchinson.)

2. 'Knees' of bald cypress (*Taxodium distichum*) in a north Florida swamp forest.

3. Vegetation destruction by sulphur dioxide in smelter fumes, Sudbury, Ontario. This denuded landscape once supported a hemlock-pine-maple forest.

4. Tank epiphytes (Bromeliaceae family) on a remnant rainforest tree, Belize.

5 & 6. A loblolly pine (*Pinus taeda*) woodland in north Florida, part of which has been protected from fire for the past thirty years (upper), and part burned annually for the last nine years (lower). Controlled fires are now widely used in vegetation management.

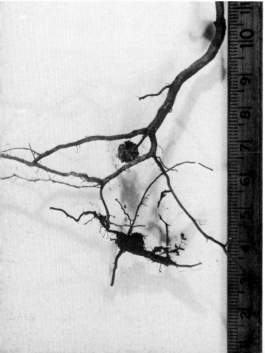

7. Root nodules containing nitrogen-fixing bacteria on the roots of red alder (*Alnus rubra*) seedlings. Pure stands of this species in the Pacific Northwest can contribute up to 70 lbs/ac/yr of nitrogen to the soil beneath them (Tarrant *et al.* 1969). (Photograph by Ian Hutchinson.)

8. The inflorescences of an *Arceuthobium* (Loranthaceae family) sprouting from the stem of its host, a Ponderosa pine (*Pinus ponderosa*) sapling, eastern Washington. These parasites are a serious forestry problem in western North America.

9. Layering by stems of vine maple (*Acer circinatum*) in a British Columbia coastal forest. When in contact with the ground and covered by litter, the long trailing branches of this species develop adventitious roots and vertical shoots. Eventually, a new independent plant can develop. (Photograph by Ian Hutchinson.)

10. Dispersion pattern of loblolly pine (*Pinus taeda*) saplings in an abandoned pasture, central Florida. High sapling densities occur adjacent to parent seed trees with a progressive diminution at increasing distance from these.

11.   Vegetation recovery after destruction by sulphur dioxide in smelter fumes, Columbia River Valley near Trail, British Columbia. Sulphur dioxide emission controls were instituted fifty years prior to this photograph. Many of the birch (*Betula papyrifera*) in the foreground appear to have regenerated from remnant root stocks. (Photograph by Bill Archibold.)

12.   Sea oats (*Uniola paniculata*) stabilizing a beach foredune, northeast Florida. The ability to withstand partial burial and to root adventitiously from buried culms is characteristic of pioneer species in sand dune successions.

B6